Motion in Biological Systems

Motion in Biological Systems

Max A. Lauffer
Andrew Mellon Professor Emeritus of Biophysics
University of Pittsburgh
Pittsburgh, Pennsylvania

Alan R. Liss, Inc., New York

Address all Inquiries to the Publisher
Alan R. Liss, Inc., 41 East 11th Street, New York, NY 10003

Printed in the United States of America

Library of Congress Cataloging-in-Publication Data

Lauffer, Max A. (Max Augustus), 1914–
Motion in biological systems / Max A. Lauffer.
p. cm.
Includes index.
ISBN 0-8451-4261-5
1. Biophysics. 2. Chemistry, Physical organic. 3. Biological systems—Mathematical models. 4. Biomechanics—Mathematical models. 5. Biological transport—Mathematical models. I. Title.
QH505.L26 1988
574.19'1—dc19 88-8278
CIP

Contents

Preface
Max A. Lauffer . . . . **ix**

List of Symbols . . . . **xi**

Chapter 1
OSMOTIC PRESSURE **1**
Thermodynamic Derivation of Osmotic Pressure Equation **2**
Excluded Volume **5**
Donnan Equilibrium **9**
Hydration of Macromolecules **14**
The Total Equation **15**
References **18**
Notes **18**

Chapter 2
FRICTIONAL RESISTANCE **23**
Viscosity of Liquids **24**
Flow Through a Capillary Tube **27**
Rotational Friction **31**
Stokes' Law **35**
Friction Coefficient of Ellipsoids of Revolution **38**
Viscosity of a Suspension of Spheres **41**
Intrinsic Viscosity of Ellipsoids of Revolution **47**
Appropriate Viscosity in Sedimentation and Diffusion **48**
Non-Newtonian Systems **50**
The Electroviscous Effect **51**
Resistance in Obstructed Systems **51**
References **53**
Notes **54**

Chapter 3
DIFFUSION **59**
Fick's Law **59**
Diffusion Through Pores **62**

Diffusion in Gels 64
Solutions of Fick's Second Law 66
Radially Symmetrical Spherical Diffusion 70
Radially Symmetrical Cylindrical Diffusion 74
Concentration-Dependent Diffusion 79
Diffusion of Charged Macromolecules 80
Thermal Diffusion 81
References 84
Notes 84

Chapter 4
MOTION IN ELECTRIC FIELDS **89**

Boltzmann Distribution Law 92
Poisson's Equation 92
Debye-Hückel Theory 94
Activity Coefficient of Electrolytes 99
Conductivity of Electrolyte Solutions 103
Electrophoretic Effect 105
Time of Relaxation Effect 108
Ionic Mobility 108
Diffusion of Salts 111
References 114
Notes 114

Chapter 5
ELECTROKINETIC PHENOMENA **115**

Theory of Electroosmosis 116
Streaming Potential 121
Electrophoresis 126
Charge Per Unit Area of a Sphere 130
Charge Per Unit Area of a Flat Surface 132
Charge on a Cylindrical Surface 133
Attempts to Verify a Theory 138
References 142
Notes 143

Chapter 6
POTENTIALS AT INTERFACES **147**

Electrode Potentials 147
Liquid Junction Potentials 152
Concentration Cells Without Transference 156
Concentration Cells With Liquid Junction 158
Salt Bridges 161

Membrane Potentials 163
Donnan Potentials 166
Resting Potentials in Biomembranes 167
References 170
Notes 171

Chapter 7
TRANSPORT ACROSS MEMBRANES 173
Anomalous Osmosis 173
Diffusion and Facilitated Diffusion 177
Thermodynamics of ATP Hydrolysis 181
Active Transport 183
Mechanisms of Permease Function 184
Transport of Macromolecules and Particles 186
References 187

Chapter 8
ENTROPY-DRIVEN PROCESSES IN BIOLOGY.
I. MECHANISM AND SIGNIFICANCE 189
Thermodynamic Background 189
Polymerization of TMV Protein 195
Significance of Entropy-Driven Processes in Biology 215
References 219
Notes 220

Chapter 9
ENTROPY-DRIVEN PROCESSES IN BIOLOGY.
II. BIOLOGICAL APPLICATIONS 221
Muscle Contraction 224
Cellular Dynamics 234
Formation of Pseudopodia in Amoebae 236
Cell Division 237
Pigment Migration 239
Cytoplasmic Streaming 239
Movement of Cilia 241
Bacterial Motility 244
References 246
Notes 247

Index 249

Preface

Biological mechanisms are enormously complex, and yet underlying each of them are a few simple physical principles. This book is addressed to dissatisfied modern biologists, those who are troubled when they cannot trace physics-based equations involved in interpretations of biological phenomena to the underlying simple physical and physico-chemical ideas. It is my hope to relieve some of this dissatisfaction.

The book consists essentially of selected topics in physics and physical chemistry of major importance in biology. There is no biology in the book not already widely known by modern biologists. When biological systems are discussed, they are discussed briefly and, therefore, incompletely and inadequately, solely for the purpose of illustrating where physical principles can be applied. In contrast, topics in physics and physical chemistry are treated in great detail. The language is, of necessity, mathematical. Familiarity with the fundamentals of algebra, geometry, trigonometry, and elementary calculus is assumed; the mathematical language in this book is confined to these. Mathematics, even elementary mathematics, does not lend itself to speed-reading. Nevertheless, I hope that the patient scholar can follow, step by step, the development of the basic ideas of physics and physical chemistry discussed in this book to the final equations so widely applied in modern biology.

Familiarity with the basic principles of Newtonian physics and with the elements of thermodynamics and kinetics usually covered in even abbreviated courses in physical chemistry is also assumed. These matters are not discussed in detail. However, some subjects, especially in the area of physical chemistry, which get omitted in some abbreviated elementary treatments, are presented in some detail. Some of the derivations of equations are merely the translation of derivations familiar to advanced students of physics and physical chemistry from the language of vector-analysis to that of elementary calculus. Several physical developments are original with the author.

I wish to acknowledge with gratitude the contributions to my thinking made by many scholar-scientists with similar interests and hopes, but especially to Professors John Edsel of Harvard, Henry Eyring of the University of Utah, and Henry Bull of the University of Iowa. In addition, I was greatly influenced in my formative years at the University of Minnesota by my mentor, Ross Aiken Gortner, and by the writings of Herbert Freundlich. All of these individuals, and many not specifically named, critically influenced me during my career, which has extended across a half century of research and teaching. It is basically out of this background that the present book emerges.

In a book containing as many equations, even basically simple equations, as this one does, opportunities for error are almost infinite. Signs can get inverted, natural logs and logs to the base ten can get mixed up, numerators and denominators can get scrambled. Worse still, all these can escape even careful proofreading. I therefore beg the indulgence of my readers and hope that they can make the obvious corrections.

I am especially grateful to my secretary of many years, Mrs. Grace Stumpf, who professes no great love of either physics or mathematics, for her patience and unusual skill in nursing the manuscript through its many changes, corrections, and revisions. I am also grateful to my daughter, Susan Keiper Lauffer, who does love mathematics, for checking the mathematical steps in the derivations and for drawing many of the illustrations. Finally, I thank, with all my heart, my wife, Erika Erskine Lauffer, for putting up with a husband who, for many years, neglected his domestic chores and looked at her with the vacant stare of the scholar-dreamer with mind far removed from the immediate scene.

Max A. Lauffer

List of Symbols

The list that follows contains the most common symbols found in numbered equations throughout the book. Some of these symbols, especially the first three and the last six letters of the alphabet, have other local meanings, defined in the text, in intermediate equations used to derive the numbered equations.

A	Avogadro's number; 6.02252×10^{23} molecules per mole
A	Excluded volume parameter
a	Minor semi-axis of ellipse
$\mathbf{a}_i$	Radius of closest approach
a	Activity, also specific value of the radius of a sphere or a cylinder
b	Major semi-axis of ellipse, also specific value of the radius of a sphere or cylinder
C	Temperature on Celsius scale
C	Integration constant
c	Concentration in various units, also semi-axis of an ellipsoid
D	Diffusion coefficient
D_T	Thermal diffusion coefficient
D	Dielectric constant
d	Diameter
E	Energy
E	Electromotive force
E	Electric intensity
e	Base of natural logarithms −2.7183
e	The electron
F	Force
F	Faraday constant; 9.64870×10^4 coulombs per mole
f	Translational friction coefficient
G	Gibbs free energy
g	Acceleration of gravity; 980.665 cm sec^{-2}
H	Enthalpy
H	Streaming potential

h Height
I Electric current in amperes
i Indefinite number
j Indefinite number
j Moles of solvent in 1000 g of solvent
K Equilibrium constant
K_a Equilibrium constant on activity basis
K_c Equilibrium constant on concentration basis
K_m Michaelis-Menten constant
K_s' Salting-out constant on ionic strength basis
K_d A constant like a salting-out constant for effect of dipolar ions on solubility
K Temperature on Kelvin scale
K Hydration factor on volume basis
k Reaction velocity constant
k Boltzmann constant; 1.38054 ergs $\mathbf{K}^{-1}$
L Length
M Molecular weight
M Moment or torque
m Molality
m Mass per molecule
N Mole fraction
ΔN Change in moles per unit volume
n Number of moles
n Number of carrier molecules in a membrane
P Hydrostatic pressure
P.E Potential energy
P Permeability
p Indicates $-\log_{10}$
Q Area
Q Heat absorbed
q Electric charge
R Gas constant; 8.31433 joules $\mathbf{K}^{-1}$ mole^{-1}; 1.9872 calories $\mathbf{K}^{-1}$ mole^{-1}
R Resistance in ohms
R Reynolds number
R_G Radius of gyration
r Radius
S Entropy
S Number of moles in diffusion equation
S The Svedberg unit of sedimentation (10^{-13} cgs units)
s Sedimentation coefficient
s An arbitrary entropy change

T	Temperature on Kelvin scale
t	Time
t	Transference number
u	Mobility of a charged particle or ion
V	Volume
V	Electric potential
v	Velocity
W	Work
X	Field strength or potential gradient
x	A Cartesian coordinate
x	Valence of a macromolecule
y	A Cartesian coordinate
z	A Cartesian coordinate
z	Valence or net charge of an ion
α	Generalized parameter in obstructed flow equations
β	Virial coefficient
$\boldsymbol{\beta}$	A concentration term
γ	Activity coefficient
Δ	Change
δ	A small constant, also a small change
ϵ	Protonic charge
$\boldsymbol{\epsilon}$	Ellipticity
ζ	Rotational friction coefficient
$\boldsymbol{\zeta}$	Zeta or electrokinetic potential
η	Viscosity coefficient
$[\eta]$	Intrinsic viscosity
θ	Angle
κ	Partition factor
$\boldsymbol{\kappa}$	Debye-Hückel constant
Λ	Equivalent conductance
$\boldsymbol{\Lambda}$	Specific conductance
λ	Equivalent conductance of an ion
$\boldsymbol{\lambda}$	Specific conductance of an ion or ionic conductance
μ	Ionic strength
$\boldsymbol{\mu}$	Chemical potential
ν	Number of molecules
$\boldsymbol{\nu}$	Number of equivalents of ions per mole of an electrolyte
ξ	Number of moles of H^+ bound per mole of protein
$\boldsymbol{\xi}$	A hydration factor
Π	Osmotic pressure
$\boldsymbol{\Pi}$	An operator designating multiplication
π	3.1416

ρ	Density
$\boldsymbol{\rho}$	Charge density
Σ	An operator indicating summation
σ	Electric charge per unit area
τ	Tangent or tangential
Φ	Volume fraction
ϕ	An angle
Ψ	Potential in the region of a charged particle
ψ	An angle
Ω	Angular velocity of a solid
ω	Angular velocity in a liquid
∇	Vector differential operator
eu	Entropy unit in calories per mole per degree **K**
f($\boldsymbol{\kappa}$r)	Function of $\boldsymbol{\kappa}$r

— above a symbol signifies per mole, or in other cases mean value, or in some equations a vector.

1

Osmotic Pressure

There are many reasons for beginning a book on movement in biological systems with a chapter on osmotic pressure. The phenomenon of osmosis, the movement of solvent into a solution when the solvent and solution are separated by a semipermeable membrane, was described in 1748 by the Abbe Nollet. He showed that when water and alcohol were separated by an animal bladder membrane, the water passed through into the alcohol, but the alcohol did not pass into the water. The plant physiologist W.F.P. Pfeffer discovered how to make an improved semipermeable membrane out of copper ferrocyanide (first used by M. Traube in 1864) and employed this method to carry out his systematic study of the osmotic pressure of various solutions. While osmotic pressure is defined operationally as the pressure that will just prevent osmosis, it is entirely dependent upon the solute concentrations in a solution. These concentrations exist whether or not there is any semipermeable membrane or any actual osmosis. Thus osmotic pressure can be considered to be a property of a solution dependent upon its solute concentrations, with a magnitude equal to the hydrostatic pressure that would be necessary to prevent osmosis if that solution were separated from solvent by a semipermeable membrane. This is the definition of osmotic pressure most commonly employed.

Pfeffer's experiments were published in 1877. His data on sucrose are of historical importance because they were used by J.H. van't Hoff in 1886 as the basis for his theory of solutions. The botanist Hugo De Vries interpreted the turgor of plant cells in terms of the movement of water from a solution of low osmotic pressure into a cell with a higher osmotic pressure and plasmolysis as the movement of water from the cytoplasm into a solution with higher osmotic pressure. Indeed, it was De Vries who called Pfeffer's work on osmotic pressure to the attention of the physical chemist van't Hoff, who then pointed out the analogy between gases and solutions. This gave van't Hoff the opportunity for the first time to apply thermodynamic methods to the study of osmotic pressure and to develop the equation used to this day for very dilute solutions relating osmotic pressure to molal concentration of solutes. H.J. Hamburger, the Dutch physiologist, published extensive experiments on hemolysis in 1883. Thus the foundation for the understanding of osmosis and osmotic presure was laid more than a century ago. A detailed historical account is presented by Glasstone [1].

Osmotic pressure can be understood in detail in terms of widely known, simple thermodynamic relationships. The development of detailed osmotic presure theory provides an opportunity to discuss the phenomenon of the excluded volume. Because of the intricate network of microtubules and other fibrous structures found within the living cell, this concept, as will be seen later, has great relevance for the intracellular movement of solutes. Finally, as will be demonstrated in a later chapter, the driving force for diffusion is the osmotic pressure gradient.

THERMODYNAMIC DERIVATION OF OSMOTIC PRESSURE EQUATION

Transfer 1 mole of pure solvent, 1, from vessel II, where $a_1'' = 1$ and the pressure is P″, isothermally to an infinite volume of a binary solution of 2 with activity a_2' and activity coefficient γ_2' in vessel I at a pressure P′, as illustrated in Figure 1.1. The symbol μ is defined as the chemical potential. At constant temperature,

$$d\mu = \frac{\partial \mu}{\partial P} dP + \frac{\partial \mu}{\partial \ln a_1} d \ln a_1 = \bar{V}_1 \, dP + RT \, d \ln a_1 = 0 \text{ at equilibrium} \tag{1.1}$$

When N_1 is the mole fraction, $\bar{V}_1$ is the molar volume and γ_1 is the activity coefficient of component 1, solvent,

$$dP = -\frac{RT}{\bar{V}_1} d \ln a_1 = -\frac{RT}{\bar{V}_1} d \ln N_1 - \frac{RT}{\bar{V}_1} d \ln \gamma_1 \tag{1.2}$$

When n_1 and n_2 are the number of moles of solvent and solute,

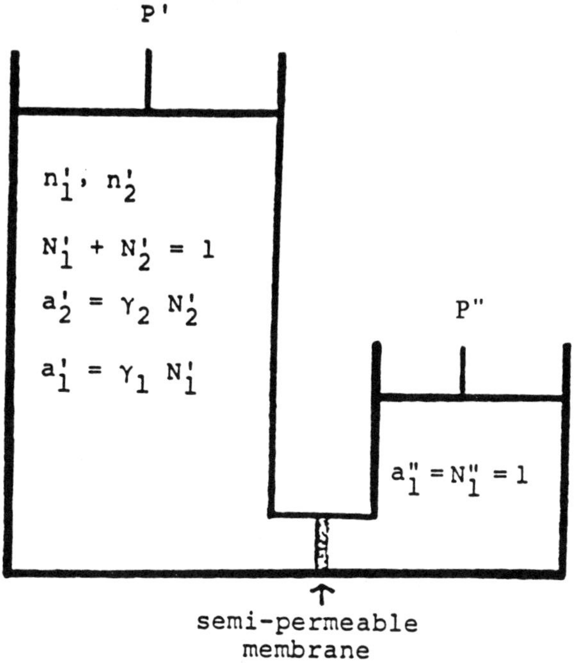

Fig. 1.1. Schematic apparatus illustrating osmotic equilibrium. Vessel I containing solvent, component 1, and solute, component 2, is separated by a semi-permeable membrane, permeable to solvent only, from vessel II, containing only solvent. Both vessels are fitted with pistons to which pressures P′ and P″ can be applied and adjusted to establish equilibrium.

respectively, from the Gibbs-Duhem equations[1], $-d \ln \gamma_1 = \frac{n_2}{n_1} d \ln \gamma_2$

$$dP = -\frac{RT}{\overline{\overline{V}}_1} d \ln N_1 + \frac{n_2}{n_1}\frac{RT}{\overline{\overline{V}}_1} d \ln \gamma_2$$

Since

$$\frac{n_2}{n_1\overline{\overline{V}}_1} = \frac{\rho_1 m_2}{1{,}000}$$

where ρ_1 is the density of the solvent and m_2 is the molality of 2,

$$dP = -\frac{RT}{\overline{\overline{V}}_1} d \ln N_1 + \frac{\rho_1 m_2 RT}{1{,}000} d \ln \gamma_2$$

$$\frac{\partial \ln \gamma_2}{\partial m_2} = \beta_{22}^{o} + \beta_{222}^{o} m_2 + \text{etc.}^2 \tag{1.3}$$

$$m_2 \, d \ln \gamma_2 = \beta_{22}^{o} m_2 \, d\, m_2 + \beta_{222}^{o} m_2^2 \, d\, m_2 + \cdots$$

$$dP = -\frac{RT}{\overline{\overline{V}}_1} d \ln N_1 + \frac{\rho_1 RT}{1{,}000} (\beta_{22}^{o} m_2 \, dm_2 + \beta_{222}^{o} m_2^2 \, dm_2 + \cdots)$$

This equation is integrated between final solution I and initial solution II.

$$\text{Since } N_1'' = 1, \Pi \equiv (P' - P'') = -\frac{RT}{\overline{\overline{V}}_1} \ln N_1' +$$

$$\frac{\rho_1 RT}{1{,}000} \left(\beta_{22}^{o} \frac{m_2^2}{2} + \beta_{222}^{o} \frac{m_2^3}{3} + \cdots \right)$$

$$\text{Since } N_1' + N_2' = 1, -\ln N_1' = -\ln (1 - N_2') = N_2' +$$

$$1/2 \, (N_2')^2 + \cdots \simeq N_2' = \frac{n_2'}{N_1' + n_2'} \simeq \frac{n_2}{n_1}$$

$$\therefore -\ln (1 - N_2') / \overline{V}_1 \simeq \frac{\rho_1 m_2}{1{,}000}$$

The approximation is only 1% in error when $m_2 = 1$ and approaches zero error as m_2 approaches 0.

$$\therefore \ \Pi \simeq \frac{\rho_1 RT}{1000}\left(m_2 + \frac{\beta_{22}^0}{2} m_2^2 + \cdots\right) \tag{1.4a}$$

A power series of this sort is frequently called a virial equation and the coefficients of the various terms in m are called virial coefficients. Thus $\rho_1 RT/1{,}000$ is the first virial coefficient, $\rho_1 RT/1{,}000)(\beta_{22}^0/2)$ is the second virial coefficient, etc. Rarely are data sufficiently accurate to justify more than two terms on the right.

Equation 1.4a, with only the first viral term on the right, is the equation for an osmotically ideal solution; it is essentially the equation originally derived by van't Hoff and is frequently referred to as the van't Hoff equation. Since this equation in its simplest form equates Π to molality, it demonstrates that osmotic pressure depends primarily on the number of solute molecules per unit volume of solvent. On a molal basis, the osmotic pressure is essentially independent of the size of the solute; macromolecules and ordinary small molecules have the same effect. However, on a mass per unit mass concentration basis, solutions of macromolecules have much lower osmotic pressures than ordinary molecules, simply because for a given mass the molality is much smaller. Fully ionized electrolytes have a larger osmotic pressure than equal molalities of nondissociable solutes because the individual ions are the particles that determine the osmotic pressure. Thus, molal sodium chloride, which exists as molal Na^+ and molal Cl^- ions, exhibits roughly twice the osmotic pressure of molal sucrose.

EXCLUDED VOLUME

The term involving RT d ln a_1 in equation 1.1 is derived from $\mu_i = \mu_{io} + RT \ln a_i$.
But

$$\mu_i \equiv \overline{G}_i = \overline{H}_i - T\overline{S}_i$$

At constant T, $d\mu_1 = d\overline{H}_1 - T\,d\,\overline{S}_1 \quad ; \quad \Delta\mu_1 = \Delta\overline{H}_1 - T\Delta\overline{S}_1$

Here $\Delta\mu_1$ means the change in the chemical potential of the solvent, and $\Delta\overline{H}_1$ and $\Delta\overline{S}_1$ are the changes in enthalpy and entropy when one mole of solvent is transferred to the large volume of solution in vessel I.

The first virial term in equation 1.4 comes from the entropy increase when the solution in vessel I is diluted by the addition of solvent. In the case in which solvent and solute molecules are aproximately the same in size and shape, the solution behaves like an ideal solution when $\Delta\overline{H}_1$ is zero. Thus, the second and higher virial terms in equation 1.4 derive from $\Delta\overline{H}_1$.

When the solute molecules are much larger than the solvent molecules, an additional term, known as the excluded volume term, is involved in $\Delta\overline{S}_1$. This leads to an additional term in m_2^2 in equation 1.4 and makes the solution behave like a nonideal solution even when $\Delta\overline{H}_1$ is zero. This factor is obviously important when the solute is a biological macromolecule. While a more general derivation of the excluded volume term involves a statistical mechanical treatment of the partial molal entropy of mixing [2], the following greatly simplified derivation is approximately correct for the special case that the macromolecules are spherical, and it serves to illustrate the fundamental physical principle involved in the more general case.

If the solute molecules are very small, ideally points, then every solute molecule in a solvent of volume V has the possibility of being anywhere in that volume. But when the solute molecules are very large, a portion of that volume is excluded to a given molecule. Thus the effective volume is less than V, and, since concentration is mass divided by volume, the effective concentration is greater than if the solute molecules were points.

Consider adding ν large spherical molecules to a volume V of solvent, one at a time until all have been added. The volume available to the first molecule added is V. Since, as illustrated in Figure 1.2, the center of a second molecule cannot approach the center of the first any closer than 2r, where r is the radius of a spherical molecule, it is excluded from a volume equal to $(4/3)\ \Pi\ (2r)^3$ or $8V_m$, where V_m is the volume of a molecule.

Molecule added	Volume available
1st	V
2nd	V − 8V
3rd	$V - 2(8V_m)$
4th	$V - 3(8V_m)$
*i*th	$V - (i - 1)\ 8V_m$

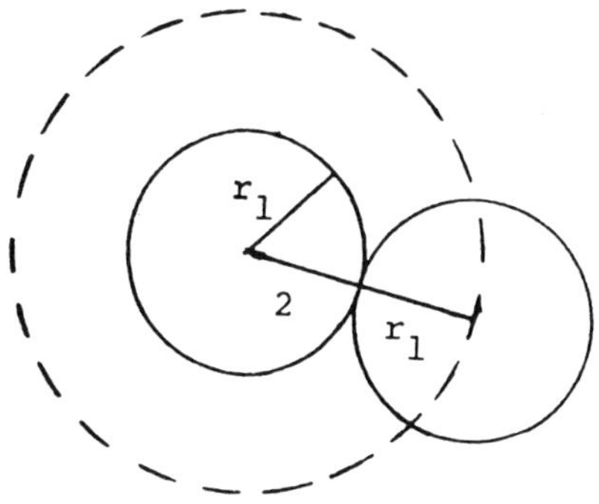

Fig. 1.2. Cross-sectional diagram of two spheres of radius r_1 in contact. The center of one sphere can approach no closer than $2r_1$ from the center of the other and is thus excluded from a volume of $8V_m$, where V_m is the volume of a single sphere.

The average volume available to the ν molecules is

$$V_{av} = \frac{\sum_1^{\nu} (V - (i - 1)\, 8V_m)}{\nu} = \frac{\nu V - \sum_1^{\nu} i(8V_m) + \nu\, 8V_m}{\nu}$$

$$= \frac{\nu V - \frac{\nu^2 + \nu}{2}(8V_m) + \nu\, 8V_m}{\nu} = V - \frac{\nu + 1}{2} 8V_m + 8V_m$$

As can be readily verified numerically

$$\sum_1^{\nu} i = \frac{\nu^2 + \nu}{2}$$

Since ν is an enormous number even in a very dilute solution, $V_{av} = V - 4\, V_m\, \nu$. The effective concentration, then, is

$$\frac{V}{V - 4\, V_m\, \nu} = \frac{1}{1 - 4\, V_m \nu / V} \simeq \left(1 + \frac{4\, V_m \nu}{V}\right)$$

times the actual concentration when $(4\, V_m \nu)/V << 1$.

Therefore, the m_2 in the *first* virial term in equation 1.4a must be replaced by

$$m_2\left(1 + \frac{4\,\nu V_m}{V}\right)$$

But

$$\frac{\nu}{V} = \frac{\mathbf{A} m_2 \rho_1}{1{,}000}$$

where m_2 is the molality, moles per 1,000 gm of solvent, **A** is Avogadro's number, and ρ_1 is the density of the solvent. If the solute molecules interact enthalpically with one another, equation 1.4 now becomes

$$\Pi = \frac{\rho_1 RT}{1{,}000}\left(m_2 + \frac{4\mathbf{A}\rho_1}{1{,}000} V_m\, m_2^2 + \frac{\beta_{22}^{0}}{2}\, m_2^2\right)$$

Frequently the excluded volume and solute interaction terms multiplying m_2^2 are combined into a newly defined $\beta_{22}^{0}/2$. For macromolecules $\Delta\overline{H}_1$ is frequently negligibly small; the interaction is then vanishingly small, and the newly defined $\beta_{22}^{0}/2$ is purely an excluded volume term. When 1,000 c/M_2 is substituted for m_2 and A_2 is defined as 4 **A** V_m/M_2^2 or $(2/3)\pi a^3/M_2^2$, where a is the diameter of the sphere, equation 1.4a can be transformed into equation 1.4b.

$$\Pi/c = \rho_1 RT(1/M_2 + \rho_1 A_2 c) \qquad (1.4b)$$

In this equation, c is concentration in grams of solute per gram of solvent. Only if concentration is defined as grams of solute per cm^3 of solvent is it dimensionally correct to omit ρ_1 in the equation. If the solute is a cylinder of length L and diameter, d, A_2 is $\pi\,\mathbf{A}\,L^2 d/4M_2^2$ [2]. As a first approximation, when the macromolecule is a randomly coiled polymer chain, A_2 is $(16/3)\pi R_G/M_2^2$, where R_G is the radius of gyration. R_G is defined as the square root of the weight average of the squares of all the distances of mass elements from the center of mass.

DONNAN EQUILIBRIUM

Donnan equilibrium also affects osmotic pressure in solutions of charged macromolecules, sometimes dramatically. In order to keep the treatment as simple as possible, the system will be assumed to be ideal. Consider a protein salt, such as K_x Pr, component 2, in a KCl solution separated from a KCl solution by a membrane permeable to K^+, Cl^-, and H_2O but not to Pr^{x-}. Ignore hydration and activity coefficients. Equation 1.1 must be obeyed by water, 1, by K^+ and Cl^-, but not by component 2. This is the criterion of Gibbs. For the ideal case,

$$\int_{P''}^{P'} \bar{V}^i \, d P + RT \int_{a_i''}^{a_i} d \ln N_i = 0$$

$$\ln \frac{N_i''}{N_i'} = \frac{\Pi}{RT} \bar{V}_i \quad ; \quad \frac{N_i''}{N_1'} = e^{\Pi \bar{V}_i/RT}$$

When separate equations are written for water, 1, K^+, and Cl^- and are then multiplied, at equilibrium,

$$\left(\frac{N_1''}{N_1'}\right)^2 \frac{N_+'N_-'}{N_+''N_-''} = \exp \frac{\Pi}{RT} (2\,\bar{V}_1 - \bar{V}_+ - \bar{V}_-) \quad ; \quad \bar{V}_+ + \bar{V}_- = \bar{V}_3$$

The exponent is extremely small for actual situations because Π/RT is, by equation 1.4a, numerically equal approximately to $m_2/1{,}000$, where m_2 is the molality of Pr^{x-}, always small.

$$\therefore \quad \left(\frac{N_1''}{N_1'}\right)^2 \frac{N_+'N_-'}{N_+''N_-''} = 1 \text{ is the condition for equilibrium.}$$

$$\frac{N'_+ N'_-}{(N'_1)^2} = \frac{N''_+ N''_-}{(N''_1)^2} \text{ at equilibrium} \tag{1.5}$$

This equation is the condition for Donnan [3] equilibrium at an osmotic presure of Π.

$$\frac{N_i}{N_1} = \qquad = \frac{n_i}{n_1} = \frac{m_i}{j}$$

$$\left(N_1 \equiv \frac{n_1}{\Sigma n_i}, \text{ etc.} \right)$$

where **j** is the number of moles of 1 in 1,000 gm of solvent.
Thus,

$$m'_+ \; m'_- = m''_+ \; m''_- \tag{1.6}$$

As a first approach, consider that vessels II and I each contain 1,000 gm of water and that they are separated by a membrane permeable to K^+ and Cl^- but not to water and Pr^{x-} ion. Initially place m_2 moles of K_xPr in vessel I and $2\overline{m}_3$ moles of KCl in vessel II, where $\overline{m}_3$ is the mean of the molalities of KCl after equilibration. During the equilibration process, δ moles of KCl will cross the membrane into vessel I. Since the m_2 moles of K_x Pr will ionize to yield $\mathbf{x}m_2$ moles of K^+ ion, m'_+ will be $\mathbf{x}m_2 + \delta$, but m'_- will be simply δ. In vessel II after equilibration, $m''_+ = m''_- = 2\overline{m}_3 - \delta$. Therefore, by equation 1.6, $(\mathbf{x}m_2 + \delta)\,\delta = (2\overline{m}_3 - \delta)^2$ at equilibrium. When this equation is multiplied out and solved for δ, one obtains

$$\delta = \frac{4\,\overline{m}_3^{\,2}}{\mathbf{x}m_2 + 4\,\overline{m}_3}$$

From equation 1.6,

$$\frac{m'_+}{m''_+} = \frac{m''_-}{m'_-} = \frac{2\,\overline{m}_3 - \delta}{\delta}$$

When the value for δ is substituted,

$$\frac{m'_+}{m''_+} = \frac{m''_-}{m'_-} = \frac{xm_2/2 + \overline{m}_3}{\overline{m}_3}$$

It should be noted that $\overline{m}_3$ is not the actual molality of KCl in either vessel but merely the mean of both.

Thus the molal concentration in vessel I of K^+ is greater than that in vessel II. There are two consequences. One is that an electromotive force will exist across the membrane. More will be said about this in later chapters.

The second consequence is that a new way of defining component 2 is suggested, a way that proves convenient for development of the osmotic pressure equation. As was done by Scatchard [4], define component 2 as K_x Pr $-$ (x/2) KCl. Start with solutions in which $n''_1 = n'_1$, $n''_3 = n'_3$ with n'_2 moles of component 2 in vessel I. Before equilibrium, in vessel II,

$$N''_1 = \frac{n''_1}{n''_1 + 2\,n''_3} \quad ; \quad N''_+ = N''_- = \frac{n''_3}{n''_1 + 2\,n''_3} \quad ; \quad \frac{N''_+ N''_-}{(N''_1)^2}$$

$$= \left(\frac{N''_3}{n''_1}\right)^2 = \left(\frac{n'_3}{n'_1}\right)^2 = \left(\frac{m'_3}{\mathbf{j}}\right)^2$$

In vessel I,

$$N'_1 = \frac{n'_1}{n'_1 + n'_2 + 2\,n'_3} \quad ; \quad N'_+ = \frac{n'_3 + \frac{\mathbf{x}}{2}\,n'_2}{n'_1 + n'_2 + 2\,n'_3} \quad ; \quad N'_- = \frac{n'_3 - \frac{\mathbf{x}}{2}\,n'_2}{n'_1 + n'_2 + 2\,n'_3}$$

$$\frac{N'_+ N'_-}{(N'_1)^2} = \left(\frac{n'_3}{n'_i}\right)^2 - \left(\frac{\mathbf{x}\,n'_2}{2\,n'_1}\right)^2 = \left(\frac{m'_3}{\mathbf{j}}\right)^2 - \left(\frac{\mathbf{x}\,m'_2}{\mathbf{j}}\right)^2 \neq \left(\frac{m'_3}{\mathbf{j}}\right)^2 = \frac{N''_+ N''_-}{(N''_1)^2} \qquad (1.7)$$

Now remove Δn_3 moles of KCl from II. $(\Delta n_3/n_1'') = (\Delta m_3/\mathbf{j})$. In order to achieve equality, Δm_3 must be so chosen that

$$\left(\frac{m_3'}{\mathbf{j}}\right)^2 - \left(\frac{\mathbf{x}\, m_2'}{\mathbf{j}}\right)^2 = \left(\frac{m_3' - \Delta m_3}{\mathbf{j}}\right)^2$$

$$(m_3')^2 - \left(\frac{\mathbf{x}}{2} m_2'\right)^2 = (m_3' - \Delta m_3)^2$$

The square root of the quantities on the left can be approximated accurately by using the binomial theorem,

$$m_3' - \frac{\mathbf{x}^2\, m_2'^2}{8\, m_3'} \simeq m_3' - \Delta m_3$$

$$\Delta m_3 = \frac{\mathbf{x}^2\, m_2'^2}{8 m_3'}$$

$$\ln \frac{N_1''}{N_1'} = -\ln \frac{n_1'' + 2n_3''}{n_1''} + \ln \frac{n_1' + 2n_3' + n_2'}{n_1'} = \ln \frac{1 + \dfrac{2n_3'}{n_1'} + \dfrac{n_2'}{n_1'}}{1 + \dfrac{2n_3''}{n_1''}}$$

$$m_3' = m_3'' + \Delta m_3$$

$$\ln \frac{N_1''}{N_1'} = \ln \frac{1 + \dfrac{2m_3''}{j} + \dfrac{m_2'}{j} + \dfrac{2\Delta m_3}{j}}{1 + \dfrac{2m_3''}{j}}$$

$$= \ln \left(1 + \frac{m_2' + 2\Delta m_3}{j\left(1 + \dfrac{2m_3''}{j}\right)} \right) \simeq \frac{m_2' + 2\Delta m_3}{j\left(1 + \dfrac{2m_3''}{j}\right)}$$

$j = 55.5$ for water and m'_2 rarely exceeds 0.01.

$$\therefore\ 1 + \frac{2m_3''}{j} \simeq 1$$

$$\therefore\ \ln \frac{N_1''}{N_1'} \simeq \frac{m_2' + 2\,\mathbf{x}^2\,(m'_2)^2/8m'_3}{j} .$$

If equation 1.2 is integrated for the ideal case in which $d \ln \gamma, = 0$,

$$\ln \frac{N_1''}{N_1'} = \frac{\Pi \bar{V}_1}{RT}$$

$$j\bar{V}_1 = 1{,}000/\rho_1$$

$$\therefore\ \Pi = \frac{\rho_1 RT}{1{,}000}\left(m_2 + \frac{\mathbf{x}^2 m_2^2}{4m_3} \right) \tag{1.8}$$

Thus, still another term is added to the second virial quantity. It is known as the Donnan term because of its derivation from Donnan equilibrium. This term is by far the largest of those in $(m_2)^2$ for most charged macromolecules.

HYDRATION OF MACROMOLECULES

There are many operational definitions of hydration applicable to macromolecules because there are many ways of determining hydration. Not all give the same answer because various models of hydration are involved in the different methods of measurement [5,6]. For thermodynamic purposes, a thermodynamic definition is required, and this definition will have the advantage of being model-free. Such a definition was proposed by Lauffer [7,8] and used by Stevens and Lauffer [9] and by Jaenicke and Lauffer [10].

If n_2 moles of *anhydrous* macromolecule are added to a vessel containing n_1 moles of solvent, 1, and n_3 moles of an ordinary solute, 3, the mole fractions of both 1 and 3 will be reduced by equal fractions because $N_i = n_i/\Sigma n_i$. If component 2 does not interact preferentially with solvent, component 1, the activities of components 1 and 3 will also be reduced by the same fraction. In actual practice, however, frequently the addition of anhydrous solute reduces the activity of water, component 1, more than that of component 3; it reduces the escaping tendency of 1 more than that of 3. This *can* be the result of an actual chemical combination of component 2 with solvent to produce a new component 2, hydrated macromolecule, thereby reducing the number of moles of free solvent and the mole fraction of free solvent. However, many other kinds of interaction can produce the same result. Be that as it may, the equality of the fractional activity reduction of components 1 and 3 can be restored by adding Δn_1 moles of water. A valid way of looking at the solution now is that it contains n_1 moles of free water, 1, n_2 moles of hydrated macromolecule, 2, hydrated to the extent of $\Delta n_1/n_2$ moles per mole, and n_3 moles of solute 3. With this method of reckoning, the fractional reduction in activity, mole fraction and, therefore, activity coefficient will be restored to equality for water and solute, 3. Hydration is defined thermodynamically as the amount of solvent, water, that must be added per mole of anhydrous solute, 2, to produce this result. When dealing with charged macromolecules and an electrolyte as component 3, it is most convenient to base the definition on the activity coefficients of solvent and solute 3.

Since $\mu 1_i = \mu 1_{io} + RT \ln a_i$ and $\Delta\mu_i = RT\,\Delta \ln a_i$, a completely equivalent way of stating the definition of hydration is that it is the amount of water that must be added along with anhydrous macromolecule to produce equal changes in μ_1 and μ_3 ; $\Delta\mu_1 = \Delta\mu_3$.

This definition of hydration is precisely equivalent to the water in a definite chemical compound of macromolecule and water that, when added to a solution of components 1 and 3, does not interact differentially with 1 and 3. Somewhat similar but not identical definitions[3] were previously proposed by Newton and Gortner [11], by Hill [12], and by Hearst and Vinograd [13].

Regardless of the mechanism by which hydration is brought about, it is the equivalent of $\Delta n_1/n'_2$ moles of water being actually chemically bound per mole of macromolecule. This means that in any solution the amount of free water is less than the total water. The molality, m'_2, of macromolecule is actually the number of moles in 1,000 gm of total water. Operationally, this is the only molality that can be measured. However, if some of the water behaves as though it were bound, the amount of free water in 1,000 gm of total water is less than 1,000 gm, and the *effective* molality of macromolecule (and of other solutes) is greater than m_2 (and m_3). If 1,000 ξ is defined as the mass of solvent bound by one mole of macromolecule, component 2, then in a solution containing 1,000 gm of solvent and m_2 moles of macromolecule, 1,000 ξm_2 is the number of grams of water bound and (1,000 − 1,000 ξm_2) or 1,000 $(1 - \xi m_2)$ is the number of grams of free water. Thus the effective molality is $m_2\,[1{,}000/1{,}000(1 - \xi m_2)] = m_2\,(1 + \xi m_2 + \xi^2 m_2{}^3$, etc.). Therefore the m_2 in the osmotic pressure equation must be multiplied by $(1 + \xi m_2$ etc.). This provides an additional quantity for the second virial term. Indeed, every m in the terms thus far derived should be multiplied by this same factor.

THE TOTAL EQUATION

When all of these terms are put together, equation 1.4a, restricted to two virial terms, becomes

$$\Pi = \frac{\rho_1 RT}{1{,}000}\left(m_2 + \left(\frac{\beta^0_{22}}{2} + \frac{4\mathbf{A}\,\rho_1 v_m}{1{,}000} + \frac{\mathbf{x}^2}{4m_3} + \xi\right) m_2{}^2\right) \tag{1.9}$$

As it appears in the above equation, β^0_{22} is derived exclusively from the enthalpic interaction of the macromolecules. Many treatments of this subject lump the enthalpic interaction and excluded volume terms in a differently defined β^0_{22} equal to their sum. Since macromolecules necessarily occur only at very low molalities, the enthalpic interaction is usually small. This is exemplified by the absence of observable temperature changes when solutions of macromolecules are diluted slightly. Thus the equation as frequently written is

$$\Pi = \frac{\rho_1 RT}{1{,}000}\left(m_2 + \left(\frac{\beta^0_{22}}{2} + \frac{\mathbf{x}^2}{4m_3} + \xi\right) m_2{}^2\right)$$

where $\beta^0_{22}/2$ is essentially equal to $4\mathbf{A}\rho_1 v_m/1{,}000$. For charged macromolecules, the Donnan term, $\mathbf{x}^2/4m_3$, is usually much the largest contributor to the second virial coefficient. When c_2 is the concentration of macromolecules in grams per gram of solvent and M_2 is its molecular weight, $m_2/1{,}000 = c_2/M_2$.

When this is substituted

$$\Pi = \rho_1 RTc_2\left(\frac{1}{M_2} + \frac{1{,}000}{M_2{}^2}\left(\xi + \frac{\beta^0_{22}}{2} + \frac{\mathbf{x}^2}{4m'_3}\right) c_2\right) \tag{1.10}$$

Expressed this way, the enormous effect the molecular weight exerts on the osmotic pressure becomes apparent.

This piecemeal derivation of the second virial coefficient is certainly not elegant, but it does focus attention one by one on the major physical consideration involved. A mathematically complete and rigorous derivation published by the author [14] leads to the same equation except for a very small correction. Unfortunately, when all of the possible interactions in a solution containing solvent, charged macromolecules, and electrolytes are carried through the derivation, the mathematics, though basically simple, becomes tedious and it is difficult to keep in mind the many physical factors involved.

For ordinary molecules with molecular weight no greater than a few hundred daltons, the first virial term is the dominant one in the osmotic pressure equation. For this reason, for many decades osmotic pressure measurements served the important function of providing reliable estimates of molecular weight. Figure 1.3, however, illustrates the enormous importance of the second virial term for solutions of charged macromolecules, and it shows dramatically the dominance of the Donnan contribution relative to others in the second virial term. This figure represents a hypothetical protein with a molecular weight of 100,000 daltons, a molar volume of 75,000 milliliters, and a charge of

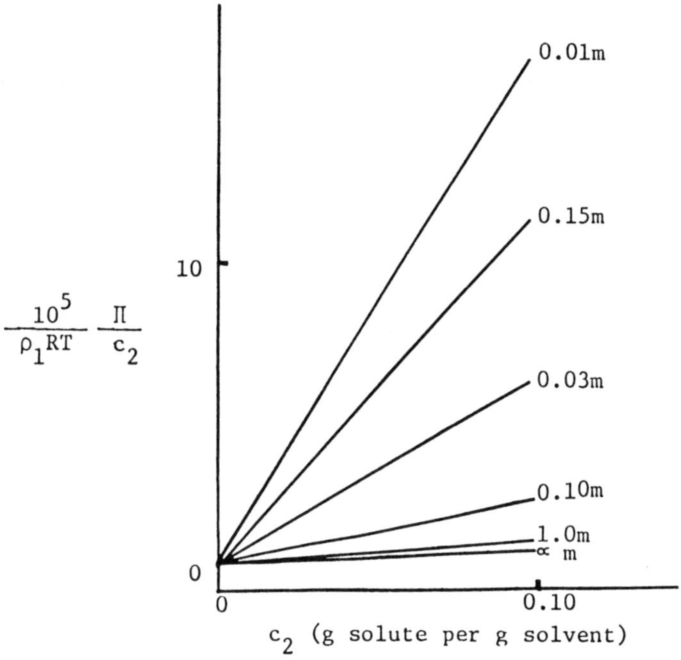

Fig. 1.3. Theoretical osmotic pressure diagram for a hypothetical protein with a molecular weight of 100,000 daltons, a molar volume of 75,000 millimeters, and a charge of −25 proton units, hydrated to the extent of 0.50 grams of water per gram of protein. $10^5\ \Pi/\rho_1\ RTc_2$ is plotted against c_2 in grams of protein per gram of solvent for various molalities of KCl. The slopes represent the second virial coefficient at the different molalities. The slope at ∞ m illustrates the small contribution made to the second virial coefficient by hydration and the excluded volume term. The steeper slopes at lower molalities of KCl illustrate the overwhelming contribution of the Donnan term.

−25 proton units in solution, hydrated to the extent of 0.50 grams of water per gram of protein. It is assumed that enthalpic protein-protein interaction is negligible. $10^5\Pi/(\rho_1 RTc_2)$ is plotted against c_2 in grams of protein per gram of solvent. The intercept at c_2 equals 0 represents the first virial term equal to the reciprocal of the molecular weight. The slopes of the various lines are measures of the second virial coefficient. The graph for ∞m KCl represents the second virial coefficient when only hydration and excluded volume are involved. The line representing 1m KCl shows that the Donnan term is very small at that electrolyte concentration. However, when the electrolyte concentration is reduced to 0.01 m KCl, the contribution of the second virual term, overwhelmingly that of the Donnan effect, is nearly 17 times that of the first virial term when the concentration of protein is 0.1 grams per gram of solvent.

REFERENCES

1. Glasstone, S. Textbook of Physical Chemistry, Second Edition. New York: D. van Nostrand Co., 1946.
2. Zimm, B. J. Chem. Phys. *14:*164, 1946.
3. Donnan, F.G. Z. Electrochem. *17:*572, 1911.
4. Scatchard, G. J. Am. Chem. Soc. *68:*2315, 1946.
5. Schachman, H.K. and Lauffer, M.A. J. Am. Chem. Soc. *72:*4266–4268, 1950.
6. Lauffer, M.A. and Bendet, I. Adv. Virus Res. *2:*241–287, 1954.
7. Lauffer, M.A. Biol. Bull. *125:*356, 1963.
8. Lauffer, M.A. Biochemistry *3:*731–736, 1964.
9. Stevens, C.L. and Lauffer, M.A. Biochemistry *4:*31–37, 1965.
10. Jaenicke, R. and Lauffer, M.A. Biochemistry *8:*3077–3092, 1969.
11. Newton, R. and Gortner, R.A. Bot. Gaz. *74:*442–446, 1922.
12. Hill, A.V. Proc. R. Soc. Lond. [Biol] *106:*477, 1930.
13. Hearst, J.E. and Vinograd, J. Proc. Natl. Acad. Sci. USA *47:*999–1025, 1961.
14. Lauffer, M.A. Biochemistry *5:*1952–1956, 1966.

NOTES

1. The Gibbs-Duhem equations, sometimes neglected in brief courses in physical chemistry, can be summarized as follows:

$$\Sigma N_i \, d \ln a_i = 0 = \Sigma n_i \, d \ln a_i$$
$$\Sigma N_i \, d \ln N_i = 0 = \Sigma n_i \, d \ln N_i$$
$$\Sigma N_i \, d \ln \gamma_i = 0 = \Sigma n_i \, d \ln \gamma_i$$

Since these equations are essential for thermodynamic reasoning, they are derived below. If the Gibbs free energy, G, of a solution is a function of T, P, n_1, n_2 etc, at constant T and P, signified by subscript T, P

$$d\,(G)_{T,P} = \Sigma \frac{\partial G}{\partial n_i} dn_i = \Sigma \mu_i \, dn_i$$

Now, applying Euler's theorem of homogeneous functions of the first degree, let $(G)_{T,P} = a\,n_1 + b\,n_2$ etc, where a and b are constants

$$\left(\frac{\partial G}{\partial n_1}\right)_{T,P} = a = \mu_1; \left(\frac{\partial G}{\partial n_2}\right)_{T,P} = b = \mu_2, \text{ etc.}$$

$$\therefore \ (G)_{T,P} = \Sigma n_i \, \mu_i$$

$$(dG)_{T,P} = \Sigma n_i \, d\, \mu_i + \Sigma \, \mu_i \, d \, n_i$$

Equate these two equations for $(dG)_{T,P}$.

$\Sigma\, n_i\, d\, \mu_i = 0$ Since $\mu_i = \mu_{io} + RTd \ln a_i$, $d\, \mu_i = RTd \ln a_i$
$\Sigma n_i\, RTd \ln a_i = 0 = \Sigma\, n_i d \ln a_i = \Sigma\, N_i d \ln a_i$
$\Sigma\, N_i = 1; \Sigma\, d\, N_i = 0 = \Sigma\, N_i d\, N_i/N_i = \Sigma\, N_i d \ln N_i = \Sigma\, n_i d \ln N_i$
$a_i = N_i \gamma_i \quad ; \quad \ln a_i = \ln N_i + \ln \gamma_i$
$\Sigma\, N_i d \ln a_i = \Sigma\, N_i d \ln N_i + \Sigma\, N_i d \ln \gamma_i = 0$
$\therefore\ \Sigma\, N_i d \ln \gamma_i = 0 = \Sigma\, n_i d \ln \gamma_i$

2. It is a matter of definition that an activity coefficient, γ_i, of a solute is 1 when the concentration of solute approaches zero and has other values at higher concentration. As long as there is a single value of γ_i for each value of m_i, this idea can be represented by the general equation:

$$\ln \gamma_i = a\, m_i + b\, m_1^2 + c\, m_i^3 \text{ etc.}$$

The equation can be made as accurate as the best available data by choosing a sufficient number of terms.

$$\beta^o_{ii} \equiv \frac{\partial \ln \gamma_i}{\partial\, m_i}_{m_i \to 0} = a \quad ; \quad \beta^o_{iii} \equiv \frac{\partial^2 \ln \gamma_i}{\partial\, m_i^{\,2}}_{m_i \to 0} = 2\, b, \text{ etc.}$$

$$\ln \gamma_i = \beta^o_{ii}\, m_i + \beta^o_{iii}\, m_i^2/2 \text{ etc.}$$
$$\partial \ln \gamma_i/\partial\, m_i = \beta^o_{ii} + \beta^o_{iii}\, m_i, \text{ etc.}$$

The operation, $\partial \ln \gamma_i/\partial m_i$, would be elegant if γ_i were defined by $a_i = \gamma_i m_i$, but, as shown by equation 1.2, γ_i was defined by $a_i = \gamma_i N_i$. The value of γ, while always unity at concentration approaching 0, at finite concentrations depends somewhat on how concentration is expressed.

The following comparison of γ' defined as a'/m with γ defined as a/N illustrates this point.

$$\mu = \mu_0 + RT \ln a = \mu_0 + RT \ln \gamma N$$

$$\mu = \mu_0' + RT \ln a' = \mu_0' + RT \ln \gamma' m$$

Since μ does not depend on how concentration is defined,

$$\Delta \mu/RT = \Delta \ln \gamma N = \Delta \ln \gamma' m$$

$\ln \gamma N/\gamma_0 N_0 = \ln \gamma' m/\gamma_0' m_0$ where the subscript, o differentiates original from final values.

$$\therefore \quad \gamma N/\gamma_0 N_0 = \gamma' m/\gamma_0' m_0$$

$$\gamma/\gamma_0 = \left(\gamma'/\gamma'_0\right)\left(m/m_0\right)\left(N_0/N\right) = \left(\gamma'/\gamma'_0\right)\left(n/n_0\right)\left(\frac{n_0}{n_s + n_0}\right) \Bigg/$$

$$\left(\frac{n}{n_s + n}\right) = \left(\gamma'/\gamma_0'\right)\left(\frac{n_s + n}{n_s + n_0}\right)$$

When n_s is the number of moles of solvent and as n_0 approaches 0, γ_0 and γ_0' approach unity,

$$\gamma = \gamma' (1 + n/n_s)$$

In a 0.1 m aqueous solution of an un-ionized solute, n is 0.1 and n_s is 55.56.

$$\gamma = \gamma' (1 + \frac{0.1}{55.56}) \; ; \quad \gamma = \gamma' (1.0018)$$

Data for biological systems are rarely if ever sufficiently accurate to distinguish between γ and γ'. Similar arguments show that this is also true when other modes of expressing concentration are used. To avoid useless complexity, the slight distinctions in biological systems between the various definitions of γ are ignored in this work.

3. Newton and Gortner (a) and Hill (b) defined bound water as total water minus the water free to dissolve an added solute with; a) a normal depression of the freezing point; or b) a normal depression of the vapor presure. Free water is thus water with the same activity as the water in the solution containing only solvent and the added solute. Hearst and Vinograd defined hydration as the number of moles of solvent that must accompany the addition of 1 mole of macromolecule to a very large volume of solution if the addition is to be at constant μ_1. All of these definitions are quantitatively close to that of Lauffer, but they neglect the thermodynamic nicety that even in a perfectly ideal system, the addition of solute 2 to a solution of 1 and 3 decreases somewhat the mole fractions, and therefore the activities, of both 1 and 3. When 2 is a macromolecule, their dilution is small.

2

Frictional Resistance

Every kind of motion except that of bodies within a vacuum is opposed by frictional forces. In biological systems, whether one is considering the movement of liquids or the movement of molecules or suspended particles through liquids, the force of resistance is related to viscosity. Active and passive transport are alike in this respect. Therefore, a consideration of the viscosity of pure liquids, especially water, of solutions, and of suspensions and the movement of molecules or suspended particles through viscous media must precede any discussion of transport. While complete treatments of this subject, when possible, involve mathematical procedures beyond the scope of this book, many relationships can be derived and discussed in the mathematical languages of algebra, geometry, and the calculus. Such simplified treatment serves the purpose of illustrating the physical principles involved.

This chapter begins with a fundamental definition of the viscosity of liquids and an attempt to understand something of the physicochemical factors contributing to viscosity. The flow of pure liquids through rigid capillaries is next considered, and, after that, the frictional resistance to the rotation of a cylinder and then of a sphere in a viscous liquid. This sets the stage for the derivation of Stokes' law for the movement of spherical particles through a viscous medium, following which is a derivation of the Einstein equation for the viscosity of a suspension of spheres; from this information, the simplest equation for the movement of fluids and particles through obstructed systems is developed. All of this can be done in a logical sequence with mathematical language no more involved than simple calculus. Even though this sequence of exercises cannot provide final answers for motion in biological systems, it will expose many of the physical principles involved. In some cases, extensions to more nearly realistic situations are presented, even though not derived fully.

In common with treatments involving other mathematical approaches, three physical assumptions underlie this development. Assumption A is that the liquid is a continuous incompressible medium. Assumption B is that any suspended particles are rigid solids large enough to be macroscopic, that is, large enough so that, by contrast, the molecular discontinuities of the solvent can be ignored. Assumption C is that there is no slippage between liquids and solid surfaces. This assumption means that the infinitesimal layer of liquid immediately adjacent to a solid surface moves with the solid. These assumptions are justified by the fact that equations derived from them accurately describe the behavior of many systems. There are exceptions, which, in some instances, can be interpreted in terms of failure of one or more of these three basic assumptions to apply to that particular case. Our treatment in this chapter is confined to systems for which these three basic assumptions are valid.

VISCOSITY OF LIQUIDS

Our starting point for the discussion of the viscosity of liquids is Newton's law of flow. This law, like all other laws of science, is a generalization that adequately describes the behavior of matter or energy under appropriate circumstances. As is illustrated in Figure 2.1, consider two parallel solid planes of area Q, separated by a distance z with a liquid of viscosity, η, between them. When the lower plane is held stationary and a horizontal force in the x-direction, F, is applied to the upper, it will move, when a steady state has been established, in the direction of the force, x, at a constant velocity, v. The liquid between the two planes can be treated as an infinite number of infinitesimally thin sheets parallel to the two planes. By assumption C, the liquid sheet adjacent to the lower plane will have a velocity of 0, and the one next to the upper plane will have a velocity of v. When the velocity is low enough to avoid turbulence, the flow of the intervening sheets will be plane laminar. Each sheet will have a slightly higher velocity than the one below it. A uniform velocity gradient, dv/dz, sometimes called the rate of shear, will be established in the liquid. For some purposes it is convenient to think in terms of shear, defined as (dv/dz) dt or simply dx/dz. For such a system, Newton's law of flow is

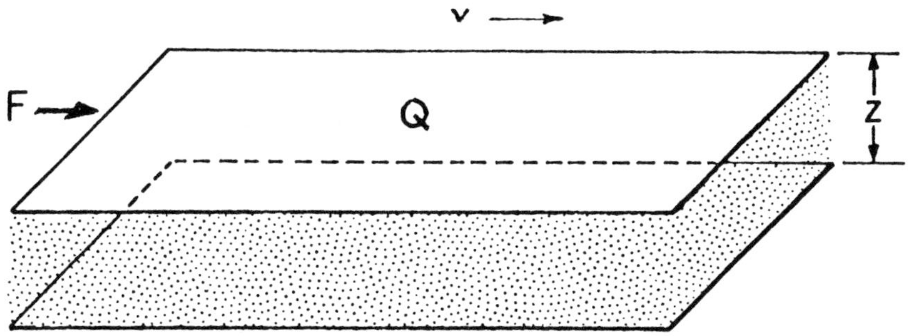

Fig. 2.1. A diagram illustrating Newton's law of flow. The space between the two parallel planes of area Q, z cm apart, is filled with liquid of viscosity coefficient, η. The lower plane is held stationary while a force F, applied in the direction of the arrow, on the upper plane causes it to move with velocity, v, in the same direction. Intervening liquid exhibits a velocity gradient, dv/dz. Reprinted [7] with the permission of the copyright owner, from the Journal of Chemical Education .

$$F = \eta Q dv/dz \qquad (2.1)$$

where η is the coefficient of viscosity with dimensions of $m^1\ l^{-1}\ t^{-1}$.

Equation 2.1 is a definition of the coefficient of viscosity. It is a law because equations deduced from it, such as equation 2.4 for the rate of flow through a capillary tube, derived in the next section, accurately describe the behavior of some liquids. Examples are water and other common laboratory solvents and solutions of small molecules and ions. Because they obey this law, such liquids are known as Newtonian liquids. Some pure liquids and also solutions of extremely anisodimensional molecules do not flow in accord with this law. They are said to be non-Newtonian.

The unit of viscosity is the poise, named after Poiseuille, about whom more will be said in the next section. A viscosity of 1 poise is that of a liquid in which a force of 1 dyne per cm^2 on the upper plane produces a velocity gradient in the liquid between the planes of 1 cm/sec cm. If the two planes such as those shown in Figure 2.1 had areas of 1 cm^2 and were 1 cm apart, the velocity of the top plane would be 1 cm/sec when there is unit velocity gradient. Thus, in one second the top plane would move 1 cm, and the work done would be 1 dyne cm or 1 erg. This corresponds to unit shear, 1 cm/cm. Thus a viscosity of 1 poise represents that of a liquid for which 1 erg/cm^3 must be expended to yield unit shear/sec. This work is all dissipated as heat. A viscosity of

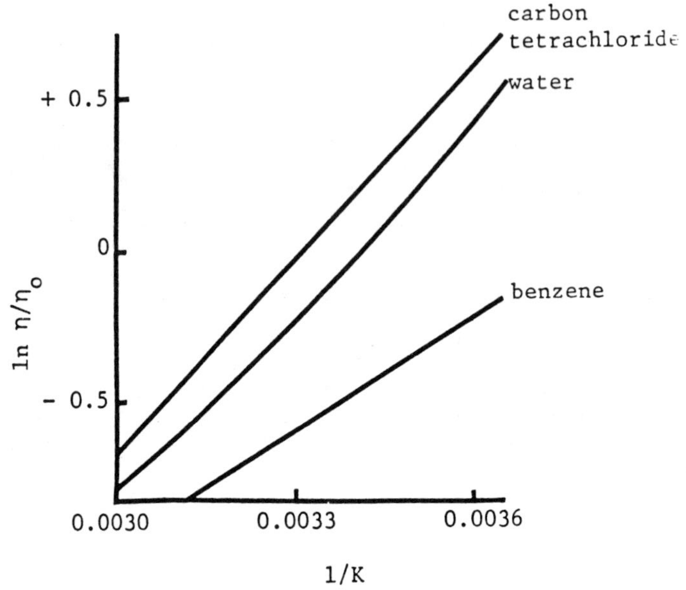

Fig. 2.2. Diagram showing ln of viscosity relative to that of water at 20°C plotted against reciprocal of temperature in **K**. The relationship is linear for the nonassociating liquids benzene and carbon tetrachloride, but deviates from linearity for the hydrogen-bonding liquid, water.

1 poise is roughly that of a medium lubricating oil at room temperature. The viscosity of water at 20°C is very close to 0.01 poise or 1 centipoise.

It was discovered early in investigations of simple liquids that the viscosity decreases as temperature is increased. It was found a long time ago that for nonassociating liquids like benzene and carbon tetrachloride the natural logarithm of the viscosity coefficient varied linearly with the reciprocal of temperature expressed as Kelvin. This is illustrated in Figure 2.2. However, as is also illustrated in that figure, the viscosity of hydrogen-bonding liquids like water deviates from this simple relationship. For the simple nonassociating liquids, the coefficient of viscosity is thus proportional to the exponential of a constant divided by RT. Arrhenius identified this constant as an energy of activation, thus relating the theory of viscosity to the theory of chemical kinetics [1]. Eyring and his associates carried this idea further [2]. The basic aspects of the Eyring point of view are that, in order for a liquid to flow, there must be: 1) regions or holes in the body of the liquid into which molecules can jump; 2) potential energy barriers that tend to prevent any particular molecule from jumping into a vacant region near it; and 3) a shearing stress to act as a mechanical potential that aids molecules jumping in the direction of the stress and hinders those jumping in the reverse direction, resulting in a net displacement in the direction of the stress. Eyring related the potential energy barrier to the heat of volatilization because the creation of a hole is in some sense related to volatilization. Based on these concepts, the view of the Eyring school is that the resultant rate of shear should be proportional to the hyperbolic sine of the ratio of the work contributed by the shearing force in moving a molecule over the energy barrier to **k**T.

The hyperbolic sine of a variable has the interesting property of being practically equal to the variable for values less than about 0.5 and practically equal to the exponential of the variable for all values greater than about 2. Thus, in a liquid in which the potential energy barrier is low, an observable rate of flow can be obtained by the application of a force small enough to be in the linear portion of the hyperbolic sine function. The rate of shear will thus be proportional to the shearing force; such a liquid will exhibit Newtonian flow. When, however, the force required to cause an observable rate of shear is significantly greater than that just mentioned, the rate of shear will not be directly porportional to the shearing force, but could be even an exponential function. Such a liquid will not exhibit Newtonian flow. But the difference between it and a Newtonian liquid is, in the view of the Eyring school, a matter of degree and not of quality.

Finally, in the Eyring view an extra term must be added for associating liquids like water. Energy must be consumed to break hydrogen bonds in addition to that required to overcome the potential energy barrier. Since the number of hydrogen bonds decreases as temperature is increased, the apparent energy of activation for viscous flow decreases as temperature increases, resulting in the curvature exhibited by water in Figure 2.2.

FLOW THROUGH A CAPILLARY TUBE

The French physician of the early 19th century, Jean-Léon Marie Poiseuille, was interested in the flow of blood. Surely he realized that blood is a complex solution and suspension and that its flow, even in the arterial system, was through branched, tapering, elastic tubes motivated by pulsating pressure. Nevertheless, he exhibited sound physical intu-

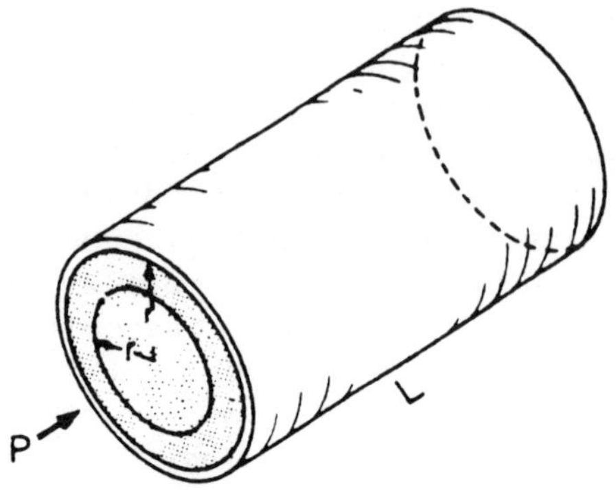

Fig. 2.3. A diagram illustrating flow of liquid of viscosity coefficient, η, through a tube of length, L, and radius, r_1, when pressure, P, is applied to one end. Reprinted [7] with the permission of the copyright owner, from the Journal of Chemical Education.

ition when he designed experiments to measure rates of flow of simple liquids through rigid capillaries of uniform bore under constant pressure difference across the tube. By measuring the volume per unit time of many different liquids in capillaries of various lengths and radii under various pressure heads, he arrived at the law that the volume of flow per unit of time was directly proportional to the fourth power of the radius and to the first power of the pressure and inversely proportional to the length [3]. Furthermore, the proportionality constant varied from liquid to liquid. Later Hagenbach [4] derived Poiseuille's law from Newton's law of flow. This derivation is available in many sources, for example, in Tanford [5], Bull [6], and Lauffer [7]. Some of the derivations differ slightly from that given below.

Assume that the liquid in a capillary tube of radius r_1 and length L consists of an infinite number of concentric cylindrical shells of radius r and thickness dr and length L with r having all values between 0 at the center and r_1 adjacent to the walls of the tube. This can be visualized as a nest of cork borers. Assume further that flow is plane laminar when a pressure P higher than the pressure at the exit end is applied. Because of assumption C, the thin cylinder immediately adjacent to the capillary wall will not move, but as one proceeds toward the center each hollow cylinder will move at an increasingly higher rate. Consider a single such hollow cylinder of radius r and thickness dr illustrated by the central solid line of Figure 2.3. This hollow cylinder will be subjected to three forces. There will be a pressure force equal to the pressure times the exposed area of the end of the cylinder, $2\pi rdr\ P$. In addition, there will be shearing forces on the inner surface and on the outer surface, because this cylinder will be moving more slowly than the one immediately inside it and more rapidly than the one immediately outside it. These forces can be calculated from Newton's law of flow, equation 2.1, where dv/dz is replaced by dv/dr and $Q = 2\pi rL$. The force on the inner surface, F', is thus $2\pi rL\eta dv/dr$. The force on the outer surface, F'', will be $2\pi L\eta[rdv/dr + (d/dr)(rdv/dr)dr]$. The net force of retardation, $F'' - F'$, will equal $2\pi L\eta(d/dr)(rdv/dr)dr$. When steady flow is achieved, this net force of retardation plus the pressure force must equal 0.

$$2\pi\ rdr\ P + 2\pi\ L\eta(d/dr)(rdv/dr)dr = 0 \tag{2.2}$$

$$-Pr/\eta L = (d/dr)(rdv/dr)$$

$$-Pr^2/2\eta L + C_1 = r(dv/dr)$$

$$v = -\ Pr^2/4\eta L + C_1 \ln r + C_2$$

Since, to satisfy the physical reality that the overall velocity of flow is always finite, v must be finite even when r is 0, C_1 must be 0.

$$v = C_2 - Pr^2/4\eta L$$

Since v is 0 when $r = r_1$, $v = P(r_1^2 - r^2)/4\eta L$. It is evident from this derivation that

$$dv/dr = -\ Pr/2\eta L \tag{2.3}$$

The velocity gradient varies from 0 in the center of the tube where r is 0 to its maximum negative value when $r = r_1$. The volume of liquid that flows in time t for this hollow shell, $V_r = v2\pi\ rdr\ t$, or

$$V = \int_0^{r1} v2\pi\ rdr\ t = \pi\ Pt\ r_1^4/8\eta L \tag{2.4}$$

Equation 2.4 is Poiseuille's law. It is validated by extensive experimentation for Newtonian liquids in which the flow rate is small enough so that the flow is plane laminar. The law breaks down when the flow rate is great enough for turbulence to set in. When the velocity becomes high enough, or when obstacles are in the path, vortices appear and large masses of fluid move as a unit, masses that possess both velocity in the direction of flow and rotational velocity. A criterion known as a Reynolds number, R, has been developed to provide a criterion for the flow velocity at which turbulence is likely to appear. This number is defined in various ways in the literature. One way of defining R for flow through a capillary tube is

$$R = 2\rho r_1^3 P/8\eta^2 L \qquad (2.5)$$

When R is defined this way, turbulence usually appears when its value exceeds 2,200. In the Reynolds equation, ρ means the density of the liquid. Non-Newtonian liquids do not obey Poiseuille's law because the apparent viscosity coefficient depends upon the velocity gradient, which, as shown by equation 2.3, varies from 0 at the center of the capillary tube to the maximum negative value at the inner walls of the capillary.

Poiseuille's law is the basis for the determination of viscosity coefficients by all methods involving flow through a capillary tube. The most commonly used method involves the Ostwald viscometer. Bingham and Jackson [8] used a highly refined capillary method involving many small correction factors to measure the viscosity of water at all temperatures at which it could be maintained as a liquid. Their data have been reproduced in many reference volumes; Table 2.1 is an abbreviated presentation. With these data as a starting point, the viscosity of any other liquid or solution can be determined by simply measuring the ratio of its viscosity to that of water at a carefully controlled temperature.

In biology one deals normally with aqueous solutions and suspensions. In general, solutes either decrease or increase the viscosity coefficient when added to water. Macromolecular solutes and suspensions of large particles generally have higher viscosities than water. Furthermore, such solutions and suspensions are frequently non-Newtonian. This question will be considered in detail in a subsequent section.

TABLE 2.1. Viscosity of Water

Temp. (°C)	Viscosity (centipoise)	Temp. (°C)	Viscosity (centipoise)	Temp. (°C)	Viscosity (centipoise)
0	1.7921	16	1.1111	32	0.7679
1	1.7313	17	1.0828	33	0.7523
2	1.6728	18	1.0559	34	0.7371
3	1.6191	19	1.0299	35	0.7225
4	1.5674	20	1.0050	36	0.7085
5	1.5188	21	0.9810	37	0.6947
6	1.4728	22	0.9579	38	0.6814
7	1.4284	23	0.9358	39	0.6685
8	1.3860	24	0.9142	40	0.6560
9	1.3462	25	0.8937	45	0.5988
10	1.3077	26	0.8737	50	0.5494
11	1.2713	27	0.8545	60	0.4688
12	1.2363	28	0.8360	70	0.4061
13	1.2028	29	0.8180	80	0.3565
14	1.1709	30	0.8007	90	0.3165
15	1.1404	31	0.7840	100	0.2838

ROTATIONAL FRICTION

Macromolecules in solution or large particles in suspension undergo constant rotational motion as a result of the kinetic energy of the system. This rotation is inhibited by friction. The inhibition can be expressed in terms of a rotational friction coefficient dependent upon the size and shape of the particle and the viscosity of the solvent or the suspending medium. The simplest case to analyze in terms of Newton's law of flow is the rotation of a cylinder of length, L, and radius, r_1, about its axis in a liquid of viscosity, η.

As illustrated in cross section in figure 2.4, imagine two concentric cylinders of length L, the inner one of radius r_1 rotating at an angular velocity of Ω radians per second and the outer one of radius r_2 being held stationary. Let the space between them be filled with a liquid of viscosity η. Ignore end effects. The liquid can be considered to be an infinite number of cylindrical shells of radius r ranging from r_1 to r_2 and of thickness dr, each rotating with its own angular velocity, ω. Because of assumption C, no slippage, the innermost at r_1 will have angular velocity $\omega = \Omega$ and the outermost at r_2, $\omega = 0$. For this case, dv/dz in equation 2.1 must be replaced by $-r\ d\omega/dr$ because $d\omega/dr$ is negative. Thus, from equation 2.1, the force on the inner surface of a cylinder at radius r is

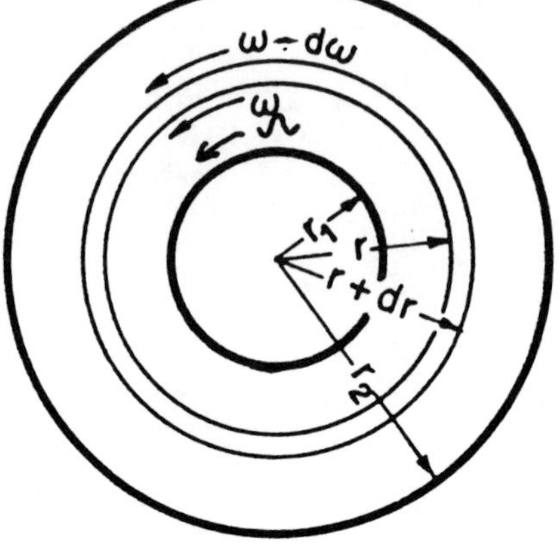

Fig. 2.4. Cross section of two concentric cylinders of length, L. The radius of the inner cylinder is r_1 and that of the outer is r_2. The space between the cylinders is filled with liquid with viscosity coefficient, η. When the inner cylinder rotates with angular velocity, Ω, radians per second and the outer cylinder is held stationary, liquids at radius, r, will rotate with angular velocity, ω. Reprinted [7] with the permission of the copyright owner, from the Journal of Chemical Education.

$$F = -\eta Q\, r\, d\omega/dr, \text{ where } Q = 2\pi r L$$

The moment or torque, **M**, is given by

$$\mathbf{M} = rF = -2\pi\, \eta L\, r^3\, d\omega/dr$$

$$\frac{\mathbf{M}}{r^3}\, dr = -2\pi\eta L d\omega$$

$$\frac{\mathbf{M}}{r^2} = 4\pi\eta L\omega + \mathbf{C}$$

When $r = r_2$, $\omega = 0$. Therefore, $\mathbf{C} = \frac{\mathbf{M}}{r^2}$. When $r = r_1$, $\omega = \Omega$

$$\mathbf{M}/r_1^2 - \mathbf{M}/r_2^2 = 4\pi\eta L\Omega \qquad (2.6)$$

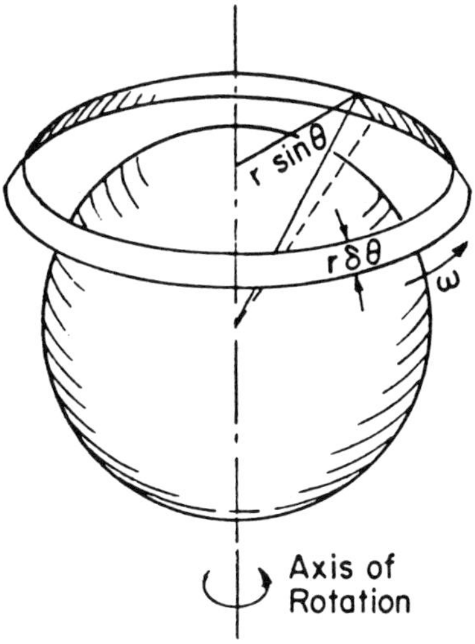

Fig. 2.5. Diagram of a sphere of radius, r_1, rotating about a vertical axis. The space outside the sphere is occupied by liquid of viscosity coefficient, η. An infinitesimally thick ring of width, $r\delta\theta$, located r cm from the center of the sphere at an angle of θ radians from the vertical direction, will also rotate about the vertical axis. The angular velocity of such a ring will vary with r, being the same as that of the sphere when r equals r_1 and zero when r approaches infinity. Reprinted [7] with permission of the copyright owner, from the Journal of Chemical Education.

$$\text{As } r_2 \rightarrow \infty,\ \mathbf{M}/r_1^2 = 4\pi\eta L\Omega$$

$$\xi \equiv \mathbf{M}/\Omega = 4\pi\eta L r_1^2 \qquad (2.7)$$

ζ is the rotational friction coefficient for rotation of a cylinder about its axis [7,9]. It is the torque required to maintain unit angular velocity with dimensions of $m^1 l^2 t^{-1}$.

With slightly more involved geometry, the same line of argument can be used to derive the equation for the rotational friction coefficient of a sphere. If a sphere of radius r_1 is rotated with angular velocity Ω about a diameter, the surrounding liquid, as illustrated in Figure 2.5, will also rotate. Consider an infinitesimally thin band of liquid at a radius r sin θ from the axis of rotation and of finite but small width, $r\delta\theta$, rotating with angular velocity ω. In this case, the velocity gradient in equation 2.1 is $-r \sin\theta\, d\omega/dr$, and the area of the shearing surface of the ring is $2\pi r \sin\theta\, r\delta\theta$. When these values are substituted into equation 2.1,

$$F_{\theta,\delta\theta} = -2\pi\eta r^3 \sin^2\theta\delta\theta\, d\omega/dr$$

$$\text{Since } \mathbf{M}_{\theta,\delta\theta} = r \sin\theta F_{\theta,\delta\theta},$$

$$\mathbf{M}_{\theta,\delta\theta}\, dr/r^4 = -2\pi\eta \sin^3\theta\delta\theta\, d\omega.$$

At the steady state, $\mathbf{M}_{\theta,\delta\theta}$ must be a constant for all values of r.

$$-\mathbf{M}_{\theta,\delta\theta}/3r^3 = -2\pi\eta\omega \sin^3\theta\delta\theta + \mathbf{C}$$

When $r \to \infty$, $\omega = 0$ and $\mathbf{C} = 0$. When $r = r_1$, $\omega = \Omega$

$$\mathbf{M}_{\theta,\delta\theta} = 6\pi\eta r_1^3\, \Omega \sin^3\theta\delta\theta \tag{2.8}$$

$$\zeta_{\theta,\delta\theta} \equiv \mathbf{M}_{\theta,\delta\theta}/\Omega = 6\pi\eta r_1^3 \sin^3\theta\delta\theta$$

$$\zeta = \sum_{\theta=0}^{\theta=\pi} \zeta_{\theta,\delta\theta} = \sum_{\theta=0}^{\theta=\pi} 6\pi\eta r^3, \sin^3\theta\delta\theta = 12\pi\eta r_1^3 \int_o^{\pi/2} \sin^3\theta d\theta = 8\pi\eta r_1^3 \tag{2.9}$$

Equation 2.9 is the Stokes equation for the rotational friction coefficient of a sphere [7,9].

The tangential velocity, v_T, of a ring $\theta\delta\theta$ on the surface of a rotating sphere is equal to r_1 ($\sin\theta$) Ω. When this is substituted into equation 2.8, one obtains

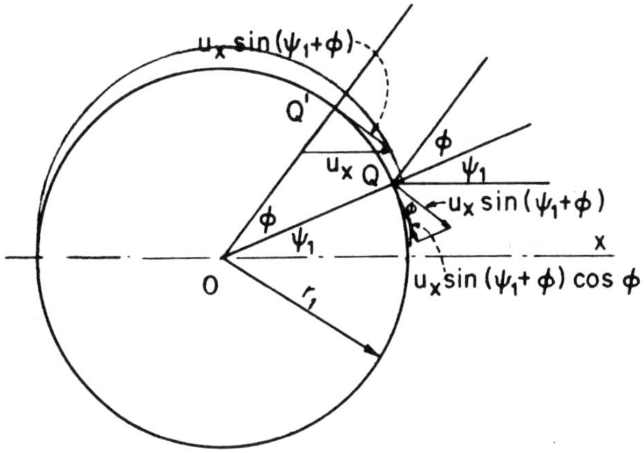

Fig. 2.6. Cross-sectional diagram of a sphere suspended in a viscous liquid moving in the x-direction with velocity, u_x. ψ_1 and $(\psi_1 + \phi)$ are angles specifying directions with respect to x. Q is a point on the surface of the sphere. The thin lines with arrows are vectors representing velocity components in the directions of the arrows. Note that in this diagram the symbol u_x is substituted for v_x as found in the text. Reprinted [7] with permission of the copyright owner, from the Journal of Chemical Education.

$$\mathbf{M}_{\theta,\delta\theta} = 6\pi\eta r_1^2 \sin^2\theta\delta\theta\, v_\tau$$

$$F = \mathbf{M}_{\theta,\delta\theta}/r_1 \sin\theta \text{ and } A = 2\pi r_1^2 \sin\theta\delta\theta$$

$$F/A = 3\eta v_\tau/r_1 \tag{2.10}$$

F/A is the shearing force per unit area exerted on the liquid at a particular position on the surface of the sphere where the tangential velocity is v_τ. This result is the starting point for determining the shearing force acting on a sphere moving with a velocity v_x in the x direction [7].

STOKES' LAW

When a sphere moves with a constant velocity, v_x, in the x direction, as illustrated in section in Figure 2.6, it is subjected to two forces, a shearing force and a pressure force. It is obvious that at $(\psi_1 + \phi) = 0$, there can be no shear, just pressure pushing against the surface, and at $(\psi_1 + \phi) = \pi/2$, there is no pressure, just shear. Because of symmetry about the x-axis, this problem can be treated as a two-dimensional problem. Results obtained for the elements of area perpendicular to the plane of the figure will be valid for all elements of area on the surface of the sphere.

Consider a point Q on the surface of the moving sphere, as illustrated in Figure 2.6. Relative to the fraction, $d\phi/2\pi$, of the liquid in the direction $(\psi_1 + \phi)$ from Q, the velocity component perpendicular to $(\psi_1 + \phi)$ is $v_x \sin(\psi_1 + \phi)$. The component of this perpendicular to OQ, the shearing component is $v_x \sin(\psi_1 + \phi)\cos\phi$, or by trigonometric rearrangement, $v_x \sin\psi_1 \cos^2\phi + v_x \cos\psi_1 \sin\phi\cos\phi$. The total shearing velocity component is the sum of the contribution from all directions of ϕ between $-\pi$ and π. Because of the symmetry mentioned above, all of these components are in the plane of the figure and are perpendicular to OQ. Thus only the magnitudes of the contributions to the shearing velocity made by fractions of the liquid $d\phi/2\pi$ need to be summed [7].

$$\begin{aligned} v_\tau &= v_x \sin\psi_1 \int_{-\pi}^{\pi} \cos^2\phi \, d\phi/2\pi + v_x \cos\psi_1 \int_{-\pi}^{\pi} \sin\phi\cos\phi/2\pi \\ &= \frac{1}{2} v_x \sin\psi_1 \end{aligned} \tag{2.11}$$

Since the equation is valid for any value of ψ_1, the subscript can be dropped. When this result is substituted into equation 2.10,

$$(F/A)_{\psi,s} = 3\eta v_x(\sin\psi)/2r_1$$

$(F/A)_{\psi,s}$ is the shearing force per unit area on the surface of the sphere at a point subtended by an angle ψ with the x-direction. This force is perpendicular to the radius OQ. Its component in the x-direction is given by

$$(F/A)_{x,s} = 3\eta v_x(\sin^2\psi)/2r_1 \tag{2.12}$$

This shearing force obviously varies from 0 at $\psi = 0$ to a maximum value at $\psi = \pi/2$. If this were the only force on the surface of the moving sphere, the liquid at the surface would slip past the surface, violating assumption C. To avoid slippage, there must be a second force, a pressure force, which varies with ψ in a way to yield a constant when added to the shearing force. This is a matter of common experience; one can actually feel a pressure when the hand is moved through water. Since for all values of ψ $\sin^2 \psi + \cos^2 \psi = 1$, this pressure force can satisfy the requirement of no slippage if

$$(F/A)_{x,p} = 3\eta v_x(\cos^2 \psi/2r_1 \quad (2.13)$$

$$(F/A)_x = 3\eta v_x(\sin^2 \psi + \cos^2 \psi)2/r_1 = 3\eta v_x/2r_1 \quad (2.14)$$

When this is multiplied by the area of a spherical surface, $4\ \pi r_1^2$

$$F = 6\pi\eta r_1\ v_x \quad (2.15)$$

The pressure and the shear equations, 2.12 and 2.13, can be integrated separately[1] to yield $F_p = 2\pi\eta r_1 v_x$ and $F_s = 4\pi\eta r_1 v_x$. As written here, the force is that exerted on the liquid in the x-direction. The force of resistance offered by the liquid to the moving sphere is equal and opposite. Equation 2.15 is Stokes's law for the frictional force on a moving sphere, derived through elementary calculus [7]. Its derivation by the method of vector analysis can be found in textbooks of theoretical physics [10].

Equation 2.15 can be written in more general terms:

$$F = fv \quad (2.16)$$

As thus written, it is understood that the force of resistance, F, is in the direction opposite the velocity. F is also the force in the direction of v required to maintain the velocity, v; f is the translational coefficient of friction with a value of $6\pi\eta r_1$ for a sphere. The dimensions of f are m^1t^{-1}.

The validity of Stokes' law was subjected to direct verification by Arnold [11]. He measured the rate of fall of small metal spheres of known density and size in oil, and then the observed rates were compared with those calculated from Stokes' law. The agreement was

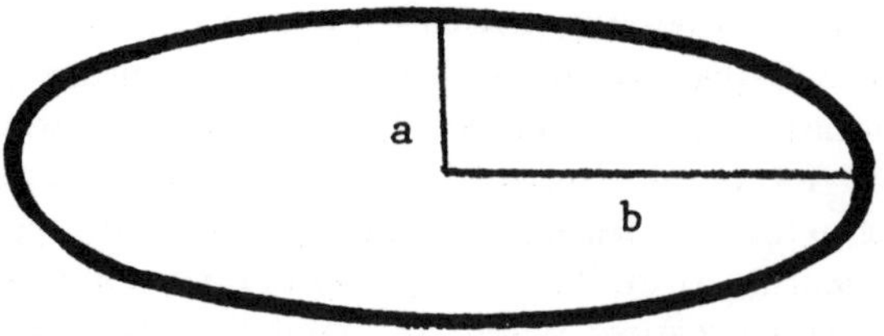

Fig. 2.7. An ellipse with major semi-axis, b, and minor semi-axis, a.

within a fraction of 1% for low velocities but less satisfactory for higher velocities. A highly sophisticated, though unintentional, verification of this law was afforded by the oil droplet experiment of Millikan [12]. In this experiment designed to determine the electronic charge, Millikan needed to know the radius of the oil droplet. This he determined from its rate of fall using Stokes' law. Determinations of the electronic charge made more recently by entirely different methods yield a value very nearly equal to Millikan's. This constitutes a highly accurate test of validity of Stokes' law.

FRICTION COEFFICIENT OF ELLIPSOIDS OF REVOLUTION

No real physical body ever fits an ideal geometric model. Nevertheless, many biological molecules and macromolecules can be treated as spheres with reasonably satisfactory results for such purposes as calculating the coefficient of friction by Stokes' law. Many others, however, are so anisodimensional that other geometric models must be sought. A highly successful model applicable to elongated particles or flattened particles is the ellipsoid of revolution. An ellipsoid of revolution is a three-dimensional geometric figure obtained by rotating an ellipse about one of its semi-axes. As illustrated in Figure 2.7, an ellipse can be characterized by a major, or long, semi-axis, b, and a minor, or short, semi-axis, a. If the a semi-axis of an ellipse is rotated about the b semi-axis, a three-dimensional figure somewhat resembling a cigar will be generated. This is called a prolate or elongated ellipsoid of revolution. Prolate ellipsoids of revolution are better models for elongated macromolecules than a sphere would be. Similarly, if the b semi-axis of the ellipse is rotated about the a semi-axis, a bun- or plate-shaped, three-dimensional figure will be generated. This is an oblate or flattened ellipsoid of revolution. It can serve as a reasonable model for many flattened or platelike particles. All ellipsoids can be characterized by three semi-axes, a, b, and c, mutually perpendicular to each other like the Cartesian coordinates in solid geometry. In a prolate ellipsoid of revolution, the a and c semi-axes are equal and are shorter than the b semi-axis. In an oblate ellipsoid of revolution, the b and c semi-axes are equal and are greater than the a semi-axis. The sphere is a special case in which the three semi-axes, a, b, c, are all equal.

Equation 2.16 can be applied to the calculation of the frictional resistance of ellipsoids of revolution. However, in this case the friction coefficient, f, depends on the kind of ellipsoid, oblate or prolate, and on the orientation of the ellipsoid with respect to the direction of motion. Beyond that, since, in the absence of restraining forces, actual particles will be randomly oriented over a very short time period, an average friction coefficient must be used. This subject is treated in three classical papers, Gans [13], Herzog et al. [14], and F. Perrin [15].

Three friction coeficients, f_a, f_b, and f_c, must be identified for each kind of ellipsoid of revolution, prolate or oblate. The subscript corresponds to the situation in which the semi-axis is pointed in the direction of the motion, that is, f_a is the coefficient of friction when the particle is oriented with its a semi-axis in line with that direction. The averaging process[2] needed to determine the f for random orientation can be given by equation 2.17.

$$\frac{1}{f} = \frac{1}{3}\left(\frac{1}{f_a} + \frac{1}{f_b} + \frac{1}{f_c}\right) \tag{2.17}$$

For prolate or elongated ellipsoids of revolution, $b > a = c$ and $f_a = f_c$. Thus, $1/f = 1/3\,[(2/f_a) + (1/f_b)]$. From the literature [13–15],

$$\text{if } \epsilon^2 \equiv \frac{b^2 - a^2}{b^2}$$

$$f_a = \frac{16\pi\eta b}{\dfrac{1}{\epsilon^2} - \dfrac{1 - 3\epsilon^2}{2\epsilon^3}\ln\left(\dfrac{1 + \epsilon}{1 - \epsilon}\right)}$$

$$f_b = \frac{16\pi\eta b}{\frac{1+\epsilon^2}{\epsilon^3}\ln\left(\frac{1+\epsilon}{1-\epsilon}\right) - \frac{2}{\epsilon^2}}$$

When these are averaged, equation 2.18 is obtained.[3]

$$f = \frac{6\pi\eta b\sqrt{1 - a^2/b^2}}{\ln\left[(b/a)(1 + \sqrt{1 - a^2/b^2})\right]} \qquad (2.18)$$

When $b >> a$,

$$f \simeq \frac{6\pi\eta b}{\ln(2\,b/a)}$$

For oblate or flattened ellipsoids of revolution,

$$f_b = f_c \text{ and } \frac{1}{f} = \frac{1}{3}\left(\frac{1}{f_a} + \frac{2}{f_b}\right)$$

$$f_a = \frac{8\pi\eta a}{\frac{1-\epsilon^2}{\epsilon^2} - (1 - 2\,\epsilon^2)\frac{\sqrt{1-\epsilon^2}}{\epsilon^3}(\arcsin \epsilon)}$$

$$f_b = \frac{16\pi\eta a}{(1 + 2\,\epsilon^2)\frac{\sqrt{1-\epsilon^2}}{\epsilon^3}(\arcsin \epsilon) - \frac{1-\epsilon^2}{\epsilon^2}}$$

These can be averaged[4] to yield equation 2.19

$$f = \frac{6\pi\eta b \sqrt{1 - a^2/b^2}}{\arcsin \sqrt{1 - a^2/b^2}} \qquad (2.19)$$

Because arc sin 1 is $\pi/2$, when $b >> a$,

$$f = 12\eta b$$

Equations 2.18 and 2.19, especially in their simplified forms when b is much greater than a, depend very strongly on the major semi-axis, b, and only slightly or not at all on the minor semi-axis, a. In the case of a thin disk, the thickness can vary widely without having any effect on f. In the case of a thin rod, considerable surface irregularity can be tolerated without greatly affecting f. One sometimes finds in the literature a version of equation 2.19 involving an arc tan instead of the arc sin. It can be shown by trigonometric rearrangement that this version is identical to equation 2.19.

VISCOSITY OF A SUSPENSION OF SPHERES

As was shown in the section of this chapter entitled Viscosity of Liquids, the viscosity coefficient, η, of a pure liquid is W, the work sec/cm^3 per unit shear. If, instead of pure liquid, a suspension of n spheres of radius r_1 is contained in 1 cm^3, the work per unit of shear will be increased. The new value of W is the sum of two terms, W_1 and W_2. W_1 derives from the fact that the spheres, being solid, will not contribute to the velocity gradient. The total volume of the spheres is $nV = \Phi$. To maintain an average velocity gradient of 1, the gradient in the liquid itself will have to be $1/(1 - \Phi)$. This will require a force per cm^2 on the "top" surface (see Fig. 2.1) of the liquid of $1/(1 - \Phi)$, and the work to displace this plane in the x direction to give unit shear, 1 cm in the x direction per cm in the z direction, will be given by equation 2.20.

$$W_1 = \eta_o/(1 - \Phi) \qquad (2.20)$$

The symbol, η_o, is used to denote the viscosity coefficient of the pure

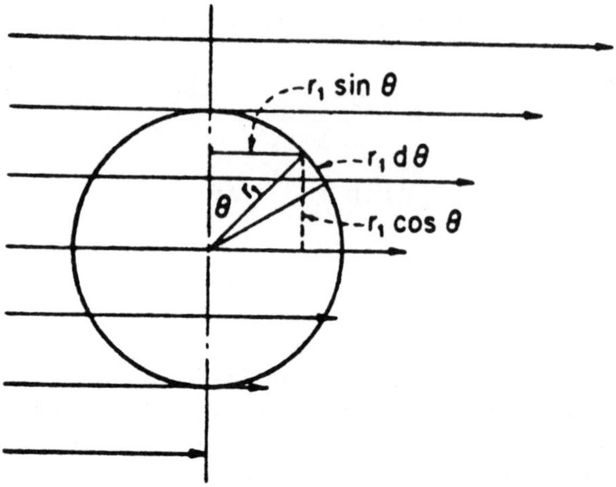

Fig. 2.8. Cross-sectional diagram of a sphere suspended in a liquid flowing with a velocity gradient. The length of each arrow represents the velocity of the liquid at the height indicated. The angle, θ, is the angle between some particular radius, r_1, and the vertical direction. Reprinted [7] with permission of the copyright owner, from the Journal of Chemical Education.

liquid or solvent in order to distinguish it from that of the whole suspension, η.

The second contribution to W derives from the fact that liquid must flow around each sphere. W_2 can be evaluated from Stokes' law. Each sphere will move with a velocity equal to that of the midplane of the flowing liquid in its immediate neighborhood. This is illustrated in Figure 2.8. When the midplane is considered to be stationary, the liquid above the sphere will move past the sphere from left to right and that below from right to left. This flow of liquid around the spheres in such a manner results in a second contribution to the work per cm^3, W_2, required to produce unit shear.

The force on a ring perpendicular to the plane of Figure 2.8 at θdθ is equal to the product of the right member of equation 2.14 and the surface area of the ring.

$$F_{\theta,x} = \frac{3\eta' v_x}{2r_1} 2\pi r_1 \sin\theta r_1\, d\theta = 3\pi r_1 \eta' v_x \sin\theta d\theta$$

Since v_x is $r_1 \cos\theta$ times the velocity gradient of the liquid,

$$v_x = r_1 \cos\theta/(1 - \Phi)$$

$$F_{\theta,x} = \frac{3\pi\eta' r_1^2 \sin\theta \cos\theta d\theta}{1 - \Phi}$$

Since the distance moved in 1 sec is $v_x = r_1 (\cos\theta/(1 - \Phi))$, the work in one second for 1 ring will be

$$\frac{3\pi\eta' r_1^3 \cos^2\theta \sin\theta d\theta}{(1 - \Phi)^2}$$

For the whole sphere

$$W_2' = 2\frac{3\pi\eta' r_1^3}{(1 - \Phi)^2}\int_0^{\pi/2} \cos^2\theta \sin\theta d\theta = \frac{2\pi\eta' r_1^3}{(1 - \Phi)^2} = \frac{1.5V\eta'}{(1 - \Phi)^2}$$

Because $nV = \Phi$, for all m spheres,

$$W_2 = \frac{1.5\Phi\eta'}{(1 - \Phi)^2} \tag{2.21}$$

In equation 2.21 and those that immediately precede it, η' means the appropriate viscosity coefficient involved in the determination of W_2.[5] At the first level of approximation, let η' be η_o, the viscosity of the pure liquid. This will be the equivalent of assuming that the spheres make no contribution to the effective viscosity, η', a condition that would be valid for very dilute suspensions. The total work sec/cm^3, equal to the viscosity, η, of the suspension is given by adding W_1 and W_2 [7].

$$\eta = \frac{\eta_o}{1 - \Phi} + \frac{1.5\eta_o\Phi}{(1 - \Phi)^2} = \eta_o\frac{(1 + 0.5\Phi)}{(1 - \Phi)^2} \tag{2.22}$$

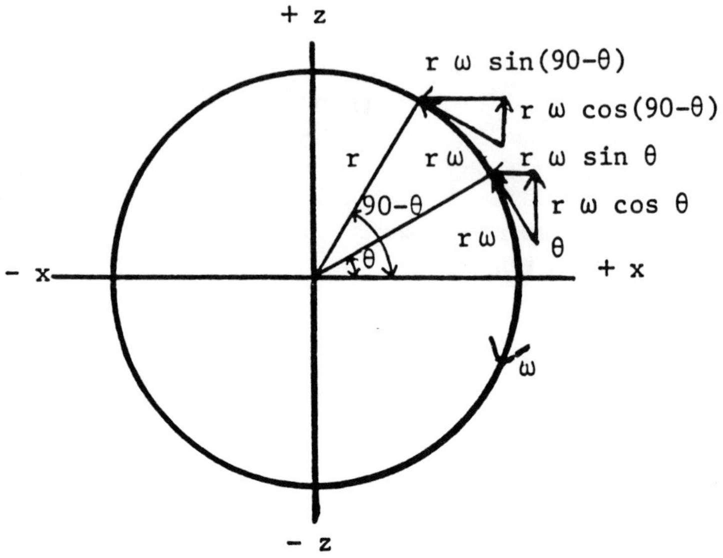

Fig. 2.9. Cross-sectional diagram representing a sphere rotating in the direction of the arrows with angular velocity ω. The vectors on the right represent the x and z components of the velocity of flow of liquid relative to that of the sphere.

This is the familiar equation originally obtained by Einstein for the viscosity of a suspension of spheres. When it is expanded by division, equation 2.22 becomes equation 2.23. Since η' in equation 2.21 was assumed to be η_o, implying that there is no contribution from other spheres, a condition that can apply only when Φ is very small, only the first two terms on the right of equation 2.23 are useful. Since r_1 does not appear in the final equation, it is valid for spheres of any size if they are large enough to be considered macroscopic. Likewise, it is valid for mixtures of spheres of different sizes.

$$\eta/\eta_o = 1 + 2.5\Phi + 4\Phi^2 + 5.5\Phi^3 \quad \text{etc.} \tag{2.23}$$

Equation 2.21 represents the work/sec/cm^3 of suspension for the liquid to move around the suspended spheres. As a first approximation in order to derive equation 2.22, it was assumed that the value of η' was that of the solvent, η_o. However, the solvent moving around each sphere is subjected to obstruction by the presence of other spheres in the suspension, and this should increase the work expended in 1 sec/cc for the liquid to move around the spheres. This means that the value of η' should be something greater than η_o. The problem then becomes one of considering how to evaluate the effect of such obstruction. Equation 2.22 can be read as a first approximation of the work/cm^3 to produce unit gradient in the suspension. However, in this case, as seen above, the velocity gradient in the liquid is greater than unity and is equal to $1/(1 - \Phi)$. Thus the work expended in 1 cm^3 to maintain unit velocity gradient of the liquid in the suspension is given by multiplying η by $(1 - \Phi)$. This is the quantity that should be used to evaluate W_2 in equations 2.21 and 2.22, because it is the obstructed liquid, not the suspension, that actually flows around each sphere.

Therefore, the value of η' in equation 2.21, instead of η_o, should be

$$\eta' = \eta(1 - \Phi) \tag{2.24}$$

Now, instead of equation 2.22, one can write

$$\eta = W_1 + W_2 = \eta_o/(1 - \Phi) + 1.5\eta\Phi(1 - \Phi)/(1 - \Phi)^2$$

$$\eta = \eta_o/(1 - 2.5\Phi), \text{ or } \eta_o/\eta = 1 - 2.5\Phi \tag{2.25}$$

The values for η/η_o given by the Einstein equation, equation 2.22, are within 1% of those given by equation 2.25 for value of Φ up to 0.05 and within 3% for Φ of 0.10, but are 25% lower for Φ of 0.25.

Equation 2.25 is here presented as a theoretical equation derived, for the first time to the knowledge of the author, from the application of Newton's law of flow, equation 2.1, to the case of spherical particles. It was proposed a long time ago by Bingham [16]. Bingham observed that for homogeneous mixtures of inert liquids, fluidities, not viscosities, are additive. Fluidity is defined as the reciprocal of the viscosity. From this experimental observation, equation 2.25 can be derived for a suspension of spheres.[6] Treffers [17] reported that the viscosity of protein solutions could best be expressed in terms of fluidity, and Bingham and Roepke [18,19] applied this concept successfully to solutions of fibrinogen and in the interpretation of the flow properties of blood serum and plasma. However, the best test of the validity of equation 2.25 is afforded by the experiments of Vand [20]. Vand made extremely careful

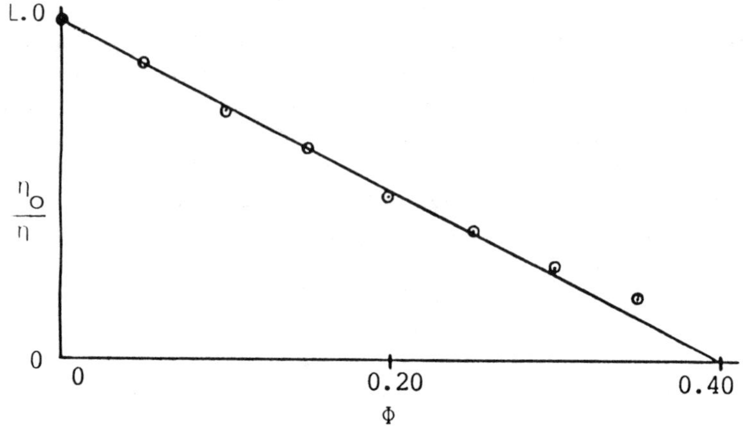

Fig. 2.10. Data of Vand [20] on the viscosity of suspensions of glass spheres in a saturated solution of zinc iodide in a mixture of water and glycerol, plotted in accordance with equation 2.25. η_o/η, the reciprocal of the specific viscosity, is plotted against the volume fraction of the spheres, ϕ.

measurements of the viscosity of suspensions of glass spheres in a saturated solution of zinc iodide in a mixture of water and glycerol. He used several different viscometers and made all conceivable experimental corrections to refine his data. The data shown in Figure 2.10 represent the average values of the relative fluidities, η_o/η, obtained in all of his viscometers, plotted against the volume fraction, ϕ. The straight line is the graph of equation 2.25. For values of Φ up to 0.25, the data fit the theoretical equation extremely well. Vand did make measurements on suspensions as concentrated as a volume fraction of 0.50, but when Φ was greater than 0.35, different results were obtained for stirred and unstirred suspensions. In any case, equation 2.25 cannot possibly apply when Φ is greater than 0.40; at that point η predicted by the equation becomes infinite, and at higher values of Φ the calculated viscosity would be negative. This cannot apply to physical reality.

As was pointed out by Bull [21] equation 2.25 can be expanded into an infinite series by simple division. $\eta/\eta_o = 1 + 2.5\Phi + 6.25\Phi^2 + 15.625\Phi^3 + \ldots$. This is not a rapidly converging series except for low values of Φ; it doesn't converge at all when Φ becomes 0.4. But from this equation, it is apparent that the limit as Φ approaches 0 of $(\eta/\eta_o - 1)/\Phi$ is 2.5 for a suspension of spheres. Kraemer [22] gave this limit the name intrinsic viscosity, usually represented by the symbol $[\eta]$. $[\eta]$ for glass

TABLE 2.2. Intrinsic Viscosity of Ellipsoids of Revolution

	$[\eta]/K$			$[\eta]/K$	
b/a	Elongated	Flattened	*b/a*	Elongated	Flattened
1.0	2.50	2.50	20.0	38.6	14.80
1.5	2.63	2.62	25.0	55.2	18.19
2.0	2.91	2.85	30.0	74.5	21.6
3.0	3.68	3.43	40.0	120.8	28.3
4.0	4.66	4.06	50.0	176.5	35.0
5.0	5.81	4.71	60.0	242.0	41.7
6.0	7.10	5.36	80.0	400.0	55.1
8.0	10.10	6.70	100.0	593.0	68.6
10.0	13.63	8.04	150.0	1222.0	102.0
12.0	17.76	9.39	200.0	2051.0	136.0
15.0	24.80	11.42	300.0	4278.0	204.0

spheres and for other unhydrated or unsolvated spheres has a value of 2.5. However, as will be shown presently, $[\eta]$ has different values for particles of other shapes. Equation 2.25 can be made more general by substituting $[\eta]$ for 2.5. When Equation 2.25 expressed in this form is expanded into a power series, it becomes $\eta/\eta_o = 1 + [\eta]\Phi + [\eta]^2\Phi^2 + \cdots$, an equation proposed by Huggins [23].

Many biological macromolecules are hydrated in solution. When particles are swollen by hydration, their effective hydrodynamic volume is $K\Phi$, where K is the ratio of the hydrated volume to the anhydrous volume. Thus, equation 2.25 for the case of spheres must be written, $\eta/\eta_o = 1/(1 - 2.5\,K\Phi)$. The intrinsic viscosity, $[\eta]$, is $2.5K$ for spherical particles. Intrinsic viscosity, $[\eta]$, depends on two properties of particles in solution or suspension, namely, the hydration factor, K, and a factor depending on particle shape equal to 2.5 for spheres, but having other values for particles of other shapes.

INTRINSIC VISCOSITY OF ELLIPSOIDS OF REVOLUTION

Simha [24] solved the problem for the case of randomly oriented elongated and flattened ellipsoids of revolution. Intrinsic viscosity depends upon the ratio of the major, b, to the minor, a, semi-axes of the ellipse whose rotation generates the ellipsoid. For elongated ellipsoids, when b/a is considerably greater than 1, equation 2.26 is obtained. For flattened ellipsoids, when b/a is considerably greater than 1, equation 2.27 is obtained. Numerical solutions are presented in Table 2.2.

$$\frac{[\eta]}{K} = \frac{(b/a)^2}{15[(\ln 2b/a) - 3/2]} + \frac{(b/a)^2}{5[(\ln 2b/a) - 1/2]} + \frac{14}{15} \qquad (2.26)$$

$$\frac{[\eta]}{K} = \frac{16}{15} \frac{b/a}{\arctan (b/a)} \qquad (2.27)$$

APPROPRIATE VISCOSITY IN SEDIMENTATION AND DIFFUSION

Equation 2.24 states that the effective viscosity of the liquid, η', in a suspension or a solution of macromolecules is the suspension or solution viscosity, η, multiplied by $(1 - \Phi)$. The remarkable success obtained when equation 2.25, derived from equation 2.24, was applied to the data of Vand establishes equation 2.24 as the correct value of viscosity to use when macromolecules in solution are moved, whatever the motivating force. This idea can be subjected to further verification by considering experiments on the sedimentation of macromolecules in the ultracentrifuge. The sedimentation coefficient, s, is the velocity of sedimentation per unit of field, whether in gravity or in the ultracentrifuge. It is inversely proportional to the viscosity. There are now strong theoretical reasons for asserting that that viscosity should be the η' of equation 2.24. However, the general practice for many years has been to assume that the viscosity of the solvent, η_o, is the proper viscosity to use. When this is done, the sedimentation coefficient, s, is found to be lower, sometimes drastically lower, when measured on concentrated solutions than the extrapolated value at infinite dilution. See the lower two curves in Figure 2.11. Lauffer [25] observed that if it is assumed that the s is inversely proportional to the solution viscosity, η, the value of s for tobacco mosaic virus is essentially independent of virus concentration. A review of the literature showed that this was also true for many other, but not quite all, macromolecules. The horizontal line at the top of Figure 2.11 shows this result for tobacco mosaic virus. The extreme deviation from the mean is $\pm 3\%$ of the mean. This is within the range of experimental error experienced in ultracentrifuge measurements at that time [26].

Lauffer [25] concluded that "the fairly general correlation between solution viscosity and the reciprocal of sedimentation rate may mean

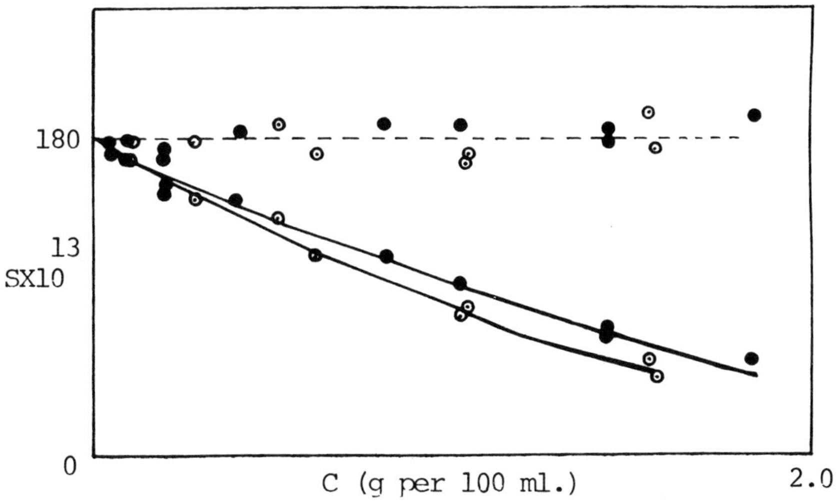

Fig. 2.11. Sedimentation coefficients in svedbergs plotted against concentration in gm/100 ml for tobacco mosaic virus. The data are those of Lauffer [25], ●, and of Schachman and Kauzmann [28], ⊙. The lower two graphs show the dependence of s on concentration when solvent viscosity is used to correct the experimental values of s and the graph at the top, fitting a horizontal line, shows that both sets of data yield s values independent of concentration when the experimental values of s are corrected for solution viscosity.

that the particles in concentrated solutions sediment through the solution instead of through the solvent." Lauffer's observation was confirmed by Miller and Price [27] with sedimentation studies on Southern bean mosaic virus and by Schachman and Kauzmann [28] with tobacco mosaic virus. However, the latter [28] concluded that the observation is "evidently merely the result of a fortuitous cancellation of two opposing effects . . ." We now know what the two opposing effects are. One of the effects was pointed out by Schachman and Kauzmann. Because a centrifuge cell is a closed system, when hydrated particles accumulate at the bottom, an equal volume of liquid must be displaced in a direction opposite that of sedimentation. In 1 second, a volume $s\Phi$ of particles will accumulate on the bottom per unit of field and of cross section. Thus a flow of liquid with a velocity equal to $s\Phi$ per unit field and unit cross section must proceed in the opposite direction. Therefore, the observed value of the sedimentation, s_{obs}, must be the true value,

$$s_{tr}, \text{ minus } s_{tr}\Phi \;\; ; \;\; s_{obs} = s_{tr}\,(1 - \Phi)$$

Since s_{tr} is proportional to $1/\eta'$, or by equation 2.24, to $1/\eta(1 - \Phi)$, s_{obs} is exactly proportional to $1/\eta$. Thus the cancellation of opposing effects can no longer be considered to be fortuitous. The idea that sedimentation in a closed system is inversely proportional to solution viscosity, not to solvent viscosity, now has strong theoretical as well as empirical support. But of even greater importance is the fact that these old ultracentrifuge results lend strong confirmation to the idea that η' of equation 2.24 is the effective viscosity for all translational motion of individual particles in a suspension of particles.

NON-NEWTONIAN SYSTEMS

Solutions and suspensions of biological materials are frequently non-Newtonian; the apparent coefficient of viscosity varies with the velocity gradient. As was pointed out earlier in this chapter, some pure liquids are non-Newtonian. However, the normal liquid in biological systems is water, which does obey Newton's law of flow. The dissolved or suspended macromolecules must be the source of the non-Newtonian behavior. One of the possible causes is hydrodynamic. The Simha equations show that the intrinsic viscosity of rodlike and platelike ellipsoids of revolution randomly oriented in the flowing stream should be much greater than the intrinsic viscosity for the same particles completely oriented parallel to the direction of flow, as given by treatments of Eisenschitz and Peterlin. Thus the intrinsic viscosity of rodlike or platelike particles will vary with the velocity gradient for purely hydrodynamic reasons, because the orientation of the particles depends upon the velocity gradient. Robinson [29] has shown that the viscosity of tobacco mosaic virus actually does vary with the degree of orientation of the particles as determined by optical means. However, the non-Newtonian behavior could be caused in part by interparticle forces, which confer upon the solution the properties of a weak gel. This effect will be called structural viscosity. Structural viscosity can be understood in terms of Eyring's concepts [2]. Eyring has postulated that a solution that shows structural viscosity is composed of two mechanical systems. The solvent and perhaps some of the dissolved particles constitute a medium that obeys Newton's law of flow. In addition, relatively few bonds of high-activation energy are assumed to exist between some of the dispersed particles, forming a network with the properties of an elastic solid. These bonds could be long-range intermolecular forces, ordinary chemical bonds, or multiple weak bonds at points of contact between dispersed particles, or forces of some unknown nature. This condition will result in a structure that will undergo only elastic deformation under low-velocity gradients, but that will flow when sufficiently high shearing forces are applied. The whole solution, consisting of the Newtonian component and the network, will exhibit non-Newtonian behavior. Eyring has shown that flow curves resembling those actually obtained for many types of system exhibiting structural viscosity can be constructed on the basis of such assumptions. Because of its anisometric solutes and suspended particles, like fibrinogen and erythrocytes, no one can expect blood to flow in a strictly Newtonian manner.

THE ELECTROVISCOUS EFFECT

When the macromolecule is charged, the viscosity of the solution or suspension is abnormally high, an effect first predicted by von Smolochowski and frequently called the electroviscous effect. Equations for this effect have been derived by Krasny-Ergen [30] and by Booth [31]. They have in common an increment proportional to the square of the electrokinetic potential and inversely proportional to the specific conductance of the solution. Kruyt and Bungenberg de Jong, Bull [32], and Briggs and his associates [33] have all verified the existence of the electroviscous effect. The intrinsic viscosity of particles with high electrokinetic potentials dissolved in media of low conductance can be many times greater than the intrinsic viscosity of the same macromolecules measured under conditions under which the charge effects are suppressed. The studies of Bull and of Briggs, however, show that neither equation affords an adequate quantitative representation of the effect of the electric charge. The results, nevertheless, do show that it can be suppressed by adjusting the pH to the isoelectric point or by adding neutral electrolytes such as sodium chloride. Electrolyte concentrations as low as 0.02 M are sufficient to reduce the effect to a relatively small value.

RESISTANCE IN OBSTRUCTED SYSTEMS

The living cell is recognized today as being something more than a bag of fluid within the cell membrane. It is an intricate structure with organelles in a network of filaments and microtubules. In many ways the cytoplasm resembles a gel. The movement of molecules or particles through the cytoplasm, whatever the motivating force, is obstructed by the network of tubules and other impenetrable bodies. It is necessary, therefore, to consider frictional resistance in obstructed systems. The simplest possible case is represented by the movement of one sphere through a fluid containing a suspension of other spheres. After this has been done, the treatment can be extended to a consideration of the movement of any particle through a system containing rodlike obstructions.

A starting point for such a consideration can be the Einstein equation, equation 2.22. As was discussed previously, this equation is reasonably accurate as long as the volume fraction of the suspended spheres is low, of the order of magnitude of 0.10 or less. The viscosity of the suspension, η by equation 2.22 is equal to $\eta_o\ (1 + 0.5\ \Phi)/(1 - \Phi)^2$. However, by equation 2.24, for reasons explained previously, the effective viscosity of the liquid in the obstructed system, η', is equal to $\eta\,(1 - \Phi)$. Thus, $\eta = \eta_o\,(1 + 0.5\,\Phi)/(1 - \Phi)$. The effective viscosity in the absence of obstruction is that of the pure liquid, η_o. This line of reasoning is contingent upon the view that it is actually the liquid in an obstructed system that interacts hydrodynamically with any particle moving with respect to the obstructed system. The way in which the obstructing solids affect the resistance encountered by the moving particle is not by direct interaction of solid with solid, but by increasing the effective viscosity of the liquid in the obstructed system. The success of this point of view, involved in the derivation of equation 2.25, in fitting the data of Vand shown in figure 2.10, lends assurance to its appropriateness. The same reasoning successfully explains the sedimentation data of figure 2.11. Since the force of resistance to a particle moving in a liquid is directly proportional to viscosity, the ratio of F′, the force of resistance of a particle moving with a given velocity in an obstructed system to F, that of the same particle moving in an unobstructed system is given by

$$F'/F = (1 + 0.5\,\Phi)/(1 - \Phi) \qquad (2.28)$$

This equation is frequently expressed in more general form as

$$F'/F = [1 + (\alpha - 1)\,\Phi]/(1 - \Phi) \qquad (2.29)$$

In this equation α has a value of 3/2 when the obstructing particles are spheres, a value of 2 when the obstructing particles are rods oriented perpendicular to the velocity gradient, and a value of 5/3 when the obstructing particles are randomly oriented rods. This obstruction effect is an old problem of 19th century physics. Important contributions to this theory were made by Lord Rayleigh [34] and more recently by Fricke [35]. This theory has been applied to diffusion in gels by Wang [36], by Lauffer [37], and by Schantz and Lauffer [38]. A limitation of the theory is the same as that of the Einstein equation from which it was derived; it is reasonably accurate only for values of Φ not much greater than 0.10.

The theory expressed in this form is the equivalent to that used by Fricke [35]. However, it would be logical to develop the theory starting with equation 2.25 instead of equation 2.22, because, as has been shown, it is much more accurate. If the same line of reasoning is followed through, one would obtain the result that $F'/F = (1 - \Phi)/(1 - 2.5\Phi)$ for spheres, or in the more general case,

$$F'/F = (1 - \Phi)/(1 - (\alpha + 1)\Phi) \qquad (2.30)$$

As far as the author knows, this equation has not been proposed previously. For a 10% suspension of spheres, equation 2.30 yields a value for F′/F of 1.2 compared with 1.17 by equation 2.29.

REFERENCES

1. Arrhenius, S. Z. Phys. Chem. *1:*285, 1887.
2. Tobolsky, A., Powell, R.E., and Eyring, H. In: The Chemistry of Large Molecules, J. Jones (ed.). New York: Interscience, 1943.
3. Poiseuille, J.L.M. Mem. Savants, E., Etrangers *9:*443, 1846.
4. Hagenbach, E. Ann. Phys. Chem. *109:*385, 1860.
5. Tanford, C. Physical Chemistry of Macromolecules. New York: John Wiley and Sons, 1961, pp. 324–336.
6. Bull, H.B. An Introduction to Physical Biochemistry, Philadelphia: F.A. Davis, 1971.
7. Lauffer, Max A. J. Chem. Ed. *58:*250–256, 1981.
8. Bingham and Jackson Bulletin, Bureau of Standards, 1475, 1918.
9. Lamb, H. Hydrodynamics, Sixth Edition. London: Dover Publications, 1928, pp. 587–589.
10. Page, L. Introduction to Theoretical Physics, Second Edition. New York: D. van Nostrand & Co., 1935, pp. 218–273.
11. Arnold, H.D. Phil. Mag. *22:*755, 1911.
12. Millikan, R.A. Pys. Rev. *32:*349, 1911.

13. Gans, R. Ann. Physik. *86:*628, 1928.
14. Herzog, R.O., Illig, R., and Kudar, H. Z. Phys. Chem. *A167:*329–342, 1933.
15. Perrin, F.J. Phys. Radium *7:*1–11, 1936.
16. Bingham, E.C. Fluidity and Plasticity, New York: McGraw-Hill, 1922.
17. Treffers, H.P. J. Am. Chem. Soc. *62:*1405, 1940.
18. Bingham, E.C. and Roepke, R.R. J. Am. Chem. Soc. *64:*1204, 1942.
19. Bingham, E.C. and Roepke, R.R. J. Gen. Physiol. *28:*79, 1944.
20. Vand, V.J. Phys. Coll. Chem. *52:*300–314, 1948.
21. Bull, H.B. Physical Biochemistry, Second Edition. New York: John Wiley and Sons, 1951.
22. Kraemer, E.O. J. Ind. Eng. Chem. *30:*1200, 1938.
23. Huggins, M.L. J. Am. Chem. Soc. *64:*2716–2718, 1942.
24. Simha, R. J. Phys. Chem. *44:*25, 1940.
25. Lauffer, M.A. J. Am. Chem. Soc. *66:*1195–1201, 1944.
26. Lauffer, M.A. and Stanley, W.M. J. Biol. Chem. *135:*463–472, 1940.
27. Miller, G.L. and Price, W.C. Arch. Biochem. *10:*467–477, 1946.
28. Schachman, H.K. and Kauzmann, W.J. J. Phys. Coll. Chem. *53:*150–162, 1949.
29. Robinson, J.R. Proc. R. Soc. Lond. [A]*170:*519, 1939.
30. Krasny-Ergen, W. Kolloid Z. *74:*172, 1936.
31. Booth, F. Proc. R. Soc. Lond. [A]*203:*533, 1950.
32. Bull, H.B. Trans. Farad. Soc. *36:*80, 1940.
33. Briggs, D.R., Hankeson, C.L., and Hanig, M. J. Phys. Chem. *45:*866 and 943, 1941 and *48:*1, 1944.
34. Lord Rayleigh, Phil. Mag. *34:*481, 1892.
35. Fricke, H. Phys. Rev. *24:*575–589, 1924.
36. Wang, J.H. J. Am. Chem. Soc. *76:*4755–4763, 1954.
37. Lauffer, M.A. Biophys. J. *1:*205–213, 1961.
38. Schantz, E.J. and Lauffer, M.A. Biochemistry *1:*658–663, 1962.

NOTES

1. F_p and F_s can be obtained by multiplying the right members of equations 2.13 and 2.12, respectively, by the area of a ring, $2\pi r_1$ (sin ψ) $r_1 d\psi$ and integrating overall values of ψ from 0 to π

$$F_p = (3\eta v_x/2r_1)(2\pi r_1^2)\int_0^{\pi} \cos^2\psi \sin\psi d\psi = 6\pi\eta v_x r_1 \int_0^{\frac{\pi}{2}} \cos^2\psi \sin\psi d\psi$$

$$F_p = 2\pi\eta r_1 v_x$$

$$F_s = (3\eta v_x/2r_1)(2\pi r_1^2)\int_0^{\pi} \sin^3\psi d\psi = 6\pi\eta\, v_x r_1 \int_0^{\frac{\pi}{2}} \sin^3\psi d\psi = 4\pi\eta\, r_1\, v_x$$

2. The problem of finding the average value of f for random

orientation can involve formidable looking equations. First, it should be noted that for a given force, F, the velocity in the direction of F is given by $F = fv$; $1/f = v/F$. Great mathematical simplification can be achieved by reducing this to a two-dimensional problem. This can be done conceptually by imagining that all of the ellipsoidal particles are aligned, for example, by a magnetic or an electrical field, so that their c semi-axes are in the z direction perpendicular to the x-y plane containing the a and b semi-axes and the directed force in the x direction, F, for example, a gravitational force. This would leave the ellipsoids free to exhibit two-dimensional random orientation within a plane. At a given instant, the a axis of the ellipsoid will make an angle θ with F and the b axis $\pi/2 - \theta$. In any time extremely small compared to the time of observation, θ will have all values between 0 and $\pi/2$ because of thermal rotary motion. Let v_a and v_b be the velocities of the ellipsoid when the force, F, is applied in the direction of the a semi-axis or the b semi-axis, respectively. The force, F, can be resolved into components parallel to a and to b to produce velocity components in the a and b direction of $v_a \cos\theta$ and $v_b \cos(\pi/2 - \theta)$, respectively. The projections of these in the direction of F are v_a and $\cos^2\theta$ and $v_b \cos^2(\pi/2 - \theta)$. The resultant velocity of the particle, v, is the sum of these.

$v = v_a \cos^2\theta + v_b \cos^2((\pi/2) - \theta)$. The average, $\bar{v}_1$ is

$$\bar{v}_1 = v_a \overline{\cos^2\theta} + v_b \overline{\cos^2((\pi/2) - \theta)} = v_a \overline{\cos^2\theta} + v_b \overline{\sin^2\theta}$$

$$\bar{v}_1 = v_a \int_0^{\frac{\pi}{2}} \cos^2\theta d\theta \Big/ \int_0^{\frac{\pi}{2}} d\theta + v_b \int_0^{\frac{\pi}{2}} \sin^2\theta d\theta \Big/ \int_0^{\frac{\pi}{2}} d\theta = \frac{1}{2}(v_a + v_b)$$

Similar equations can be derived by identical means for the average velocity, $\bar{v}_2$ and $\bar{v}_3$, respectively, when the b and the a semi-axes are restricted to the z direction

$$\bar{v}_2 = 1/2\,(v_a + v_c)$$

$$\bar{v}_3 = 1/2\,(v_b + v_c)$$

In all of these equations, the same average velocity in the x-direction is obtained whether the two free semi-axes are randomly oriented or each spends half time oriented in the x-direction.

Now suppose that in a single short time t, each semi-axis is restricted to the z direction t/3. When the c axis is restricted to the z direction, the a semi-axis will behave hydrodynamically as though it were oriented in the x direction for $t/3 \times 1/2 = t/6$. When the b semi-axis is similarly restricted, the a semi-axis will also behave hydrodynamically as though it were oriented in the x direction for t/6. Thus, in the total cycle, the a semi-axis behaves hydrodynamically as though it were oriented in the x direction for $t/6 + t/6 = t/3$. Similar reasoning leads to the result that all three semi-axes behave as though they were oriented in both the x and the y directions for t/3, and each is actually oriented in the z direction for t/3. The average velocity in the hypothetical experiment will be

$$\bar{v} = 1/3\,(\bar{v}_1 + \bar{v}_2 + \bar{v}_3) = 1/3\,[(v_{a/2} + v_{b/2}) + (v_{a/2} + v_{c/2}) + (v_{b/2} + v_{c/2})]$$

$$\bar{v} = 1/3\,(v_a + v_b + v_c)$$

Since it has been established that it doesn't matter whether the axes are oriented in a fixed direction for a fraction of the time or are randomly oriented, this equation applies for the fully random case. Since 1/f is proportional to v, for random orientation,

$$1/f = 1/3\,(1/f_a + 1/f_b + 1/f_c) \tag{2.17}$$

3.

$$48\pi\eta b/f = 2\left(\frac{1}{\epsilon} - \frac{1 - 3\epsilon^2}{2\epsilon^3}\ln\left(\frac{1+\epsilon}{1-\epsilon}\right) + \frac{1+\epsilon^2}{\epsilon^3}\ln\left(\frac{1+\epsilon}{1-\epsilon}\right) - \frac{2}{\epsilon^3}\right.$$

$$= \frac{4\ln\dfrac{1+\epsilon}{1-\epsilon}}{\epsilon}\ ; \quad \epsilon = \sqrt{-a^2/b^2}$$

$$48\pi\eta b/f = \frac{4\ln\left(\dfrac{1+\sqrt{1-a^2/b^2}}{1-\sqrt{1-a^2/b^2}}\right)}{\sqrt{1-a^2/b^2}} = \frac{4\ln\left(\dfrac{1+\sqrt{1-a^2/b^2}}{a/b}\right)^2}{\sqrt{1-a^2/b^2}}$$

$$= \frac{8\ln[(b/a)(1+\sqrt{1-a^2/b^2})}{\sqrt{1-a^2/b^2}}$$

The second term on the right was obtained by multiplying both the numerator and the denominator in ln by $(1 + \sqrt{1 - a^2/b^2})$

$$f = \frac{6\pi\eta b\sqrt{1-a^2/b^2}}{\ln[(b/a)(1+\sqrt{1-a^2/b^2})]}$$

4.

$$24\pi\eta a/f = \frac{1-\epsilon^2}{\epsilon^2} - \frac{(1-2\epsilon^2)\sqrt{1-\epsilon^2}}{\epsilon^3}\arcsin\epsilon$$

$$+ \frac{(1+2\epsilon^2)\sqrt{1-\epsilon^2}}{\epsilon^3}\arcsin\epsilon - \frac{1-\epsilon^2}{\epsilon^2}$$

$$= \frac{4\sqrt{1-\epsilon^2}}{\epsilon}\arcsin\epsilon$$

$$f = \frac{6\pi\eta a\epsilon}{\sqrt{1-\epsilon^2}\arcsin\epsilon} = \frac{6\pi\eta b\sqrt{1-a^2/b^2}}{\arcsin\sqrt{1-a^2/b^2}}$$

5. Equation 2.21 was derived for the condition that the spheres do not rotate. In fact they do rotate. As represented in Figure 2.9, each sphere will rotate in a clockwise direction about the y axis (the axis perpendicular to the plane of the paper) with some definite angular velocity, ω. An element of area at the end of a radius r that makes an angle θ with the x axis will have a peripheral velocity of r ω. This will have a component in the x-direction of r ω sin θ that will diminish the contribution to W_2' by an amount proportional to r ω sin θ. It will also have a component in the z direction that will contribute to W_2' by an amount proportional to r ω cos θ. But for every element of area at the end of a radius making an angle θ, there will be one at the end of a radius making an angle $(90 - \theta)$. For this element of area, the contribution to W_2' will be diminished by an amount proportional to r ω sin $(90 - \theta) =$ r ω cos θ and will be enhanced by an amount proportional to r ω cos $(90 - \theta) =$ r ω sin θ. Thus the diminution of the contribution to W_2' at one position, θ, resulting from rotation is exactly balanced by an added contribution at a position $(90 - \theta)$. The net effect is that the energy dissipation from the sphere as it rotates in a velocity gradient is the same as that computed without taking rotation into account.

6. Let η_o be the viscosity of the primary liquid (solvent) and η^* that of a secondary liquid (solute) at volume fraction Φ. If η is the viscosity and $1/\eta$ the fluidity of the solution, and if fluidities are additive, $1/\eta = (1 - \Phi)/\eta_o + \Phi/\eta^*$ and $\eta_o/\eta = 1 - \Phi + (\eta_o/\eta^*)\Phi = 1 - (1 - \eta_o/\eta^*)\Phi$. The term in parentheses on the right is a constant. When the solute is spherical solid, that constant is 2.5 as suggested by the Einstein equation, equation 2.22 Thus, $\eta_o/\eta = 1 - 2.5\Phi$.

3

Diffusion

FICK'S LAW

It was pointed out in Chapter 1 that an osmotic pressure gradient is the driving force in diffusion. Pressure is force per unit area and pressure gradient is force per unit volume. What actually happens in osmosis is that solvent moves from a region of lower solute activity into a region of higher solute activity. Thus an osmotic pressure gradient equals the force per unit volume acting on the solvent to move it in the direction of lower to higher solute activity. Now consider an element of volume of area, Q, and thickness dx in which there is an osmotic pressure gradient, $-d\Pi/dx$. The force tending to move solvent in the $-x$ direction will be $-(d\Pi/dx)Qdx$. In a closed vessel, the only way that solvent can move in the $-x$ direction is for solute to move in the $+x$ direction. The number of molecules of solute in Qdx is $c\mathbf{A}Qdx$, where c is now defined as moles of solute per unit of free volume (solvent plus solute). Thus the force per solute molecule is $-(d\Pi/dx)/c\mathbf{A}$.

This force will cause the solute molecules to accelerate in the x direction until their terminal velocity, v, is reached. According to equation 2.16, the force of resistance is fv, where f is the coefficient of friction proportional to the viscosity. At the steady state, $fv = -(d\Pi/dx)/c\mathbf{A}$ or $v = -(d\Pi/dx)/c\mathbf{A}f$. In time, dt, vdtQc moles of solute will pass through the area Q. This will be called d $\mathbf{S}$.

$$\therefore \quad d\mathbf{S} = -(Q/\mathbf{A}f)(d\Pi/dx)dt \qquad (3.1)$$

In equation 1.10 relating osmotic pressure, Π, to concentration of solute, c_2 is defined as grams per gram of solvent. c_2/M_2 would represent moles of solute per gram of solvent and $\rho_1 c_2/M_2$ would be moles of solute per cm^3 of solvent. This is equal to c, the moles of solutes per cm^3 of solution divided by $(1 - \Phi)$, where Φ is the volume fraction of solute. Thus equation 1.10 can be rewritten as

$$\Pi = \frac{RTc}{1 - \Phi}\left(1 + 1{,}000\left(\zeta + \frac{\beta^{o}_{22}}{2} + \frac{\mathbf{x}^2}{4m'_3}\right)\frac{c}{1 - \Phi}\right)$$

$$\text{As } c \rightarrow 0,\ \Phi \rightarrow 0 \text{ and } d\Pi/dx = RT\ dc/dx$$

When this is substituted into equation 3.1 and when R/**A** is replaced by the Boltzman constant, **k**, equation 3.2 is obtained.

$$d\,\mathbf{S} = -(\mathbf{k}T/f)\ Q\ (dc/dx)\ dt \tag{3.2}$$

Since **k**T/f for a given molecule of fixed shape depends only on T and the viscosity, equation 3.2 can also be written in the more usual form

$$d\,\mathbf{S} = -DQ(dc/dx)\ dt \tag{3.3}$$

$$D \equiv \mathbf{k}T/f \tag{3.4}$$

Equation 3.3, derived here from osmotic pressure theory, was first proposed by Fick [1] and is usually called Fick's first law of diffusion. Fick arrived at this equation by realizing that his own extensive diffusion experiments were analogous to the results for heat conduction, the mathematical equation for which was earlier derived by Fourier [2]. Equation 3.4 is the Einstein-Sutherland equation, here asserted as a definition, but originally derived to interpret diffusion. This equation is useful in showing how D varies with T and η, which in turn varies with T. It also permits the evaluation of f from measurements of D, and, from this, information about the size and shape of the solute molecule can be deduced.

A more general form of the diffusion equation, usually called Fick's second law, can be derived from equation 3.3. First, it must be realized that every concentration gradient is a vector, because is has direction as

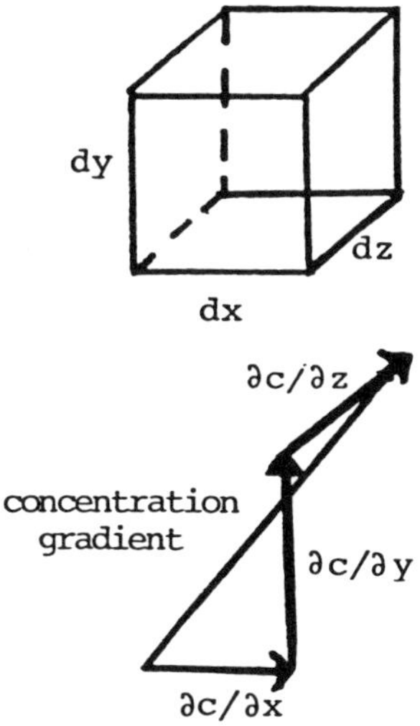

Fig. 3.1. Diagram of an elemental cube, dx dy dz (top). The vector diagram shows how the concentration gradient can be resolved into components parallel to each of the directions of the cube (bottom).

well as magnitude. Since it is a vector, it can be resolved into components parallel to the x, the y, and the z Cartesian coordinates, $\partial c/\partial x$, $\partial c/\partial y$, and $\partial c/\partial z$, respectively. Now consider an elemental cube, dx dy dz, in a region where there is a concentration gradient, oriented as in Figure 3.1. By equation 3.3, the number of moles diffusing into the cube in the x-direction from the left through the area dy dz is $-D\,dy\,dz\,\partial c/\partial x$, and the number diffusing out on the right is

$$-D\,dy\,dz\left(\frac{\partial c}{\partial x}+\frac{\partial}{\partial x}\frac{\partial c}{\partial x}\,dx\right)dt.$$

The accumulation in the cube is the difference between them, $D\,dx\,dy\,dz\,(\partial^2 c)/(\partial x^2)\,dt$. Similarly, the moles accumulated from diffusion in the y- and z-directions are, respectively, $D\,dx\,dy\,dz\,(\partial^2 c)/(\partial y^2)\,dt$ and $D\,dx\,dy\,dz\,(\partial^2 c)/(\partial z^2)\,dt$. The total number of moles accumulated divided by the volume, dx dy dz, is the change in concentration, dc. Thus

$$dc/dt = D\left(\frac{\partial^2 c}{\partial x^2} + \frac{\partial^2 c}{\partial y^2} + \frac{\partial^2 c}{\partial z^2}\right) \tag{3.5}$$

When diffusion can be limited to one direction only, for example, the x-direction, the equation is simplified by dropping the y and z terms. A more general way to write this equation is

$$dc/dt = D\,\nabla^2 c \tag{3.6}$$

where ∇^2 means

$$\left(\frac{\partial^2}{\partial x^2} + \frac{\partial^2}{\partial y^2} + \frac{\partial^2}{\partial z^2}\right)$$

in the Cartesian coordinate system, but has different meanings for systems with spherical or cylindrical coordinates. Equation 3.6, called Fick's second law, is really not a new law; because it can be derived from equation 3.3, it is merely a general aspect of Fick's law of diffusion.

DIFFUSION THROUGH PORES

Perhaps the simplest biological application of diffusion would be represented by the transport of a metabolite through a pore in a membrane. Assume that on the outside the pore is bathed by a solution of the metabolite at a constant concentration, c_0, and that on the inside of the pore the concentration is maintained at some constant value, c_i, perhaps nearly 0 because the metabolite is rapidly converted into something else. The main requirement of the pore to permit a simple deduction is that it be large enough in cross section at every point so that it does not act as a filter. It can hardly be expected that a pore in a biological membrane will be of uniform cross section or even that it will run in a single direction. However, the simplest way to begin the treatment of this problem is to consider diffusion through a straight tube of uniform cross section bathed on the outside by molecules at concentration c_0 and on the inside by molecules at constant concentration c_i. Modifications to accommodate variable cross section in variable directions can be introduced later.

When a steady state is achieved, solute molecules will not accumulate in the tube; they will merely pass through it. Thus dc/dt will equal 0 in the tube. By equation 3.5 restricted to gradients in the x-direction only, Ddc^2/dx^2 will also equal 0. Upon integration this leads to the result that dc/dx is constant. Therefore, diffusion in this system can be interpreted by equation 3.3, Fick's first law, where Q is the cross-sectional area of the tube and the constant value of dc/dx is equal to the difference between the concentrations inside and outside, Δc, divided by the length of the tube, L, and equation 3.3 can be integrated to give the result

$$\Delta S/\Delta t = -DQ\Delta c/L \qquad (3.3a)$$

In this equation, Q/L is a constant depending on the geometry of the tube and the number of moles of solute transported in time Δt is proportional to D and Δc.

Now consider the next level of complication. Suppose that in a straight tube the cross-sectional area, Q, varies from position to position along L. Since all of the material that enters the tube on the outside in unit time must be equal to all of the material that leaves it on the inside in unit time, dS/dt of equation 3.3 must be a constant everywhere in the tube. Thus, Q (dc/dx) must be a constant everywhere. In regions where Q is smaller than average, dc/dx must be greater than average, but the product must be everywhere a constant equal to $\overline{Q}\,\Delta c/L$. Finally, if the tube of nonuniform cross section is also not straight, the effect will be that the actual path length is greater than the distance between the ends of the tube, L, but it will still be a constant. Thus, for this ultimately complicated situation, which probably corresponds to real pores in membrane, the Q/L of the last equation must be replaced by another constant, $(\overline{Q/L})$, and the rate of transport across the pore is still proportional to D and Δc. Not only does this complex type of pore probably resemble a pore in a membrane, it also resembles the pores in one of the simplest apparatuses for measuring diffusion coefficients, the Northrop-Anson cell [3]. As is illustrated in Figure 3.2, this cell is nothing more than a sintered glass filter fused to a glass cup. A solution is placed inside the cup and the outside is immersed in solvent. Convection will keep the concentration in the cup immediately adjacent to the filter at its initial value and that outside the filter at the value of 0. To use the method it is first necessary to evaluate the constant

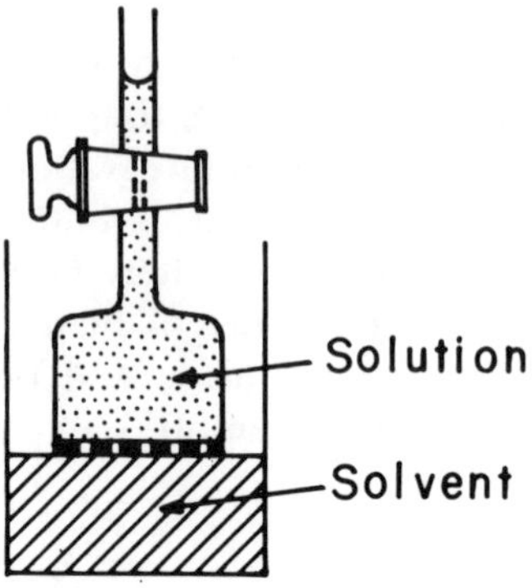

Fig. 3.2. Diagram of the porous-disk diffusion apparatus. Reprinted from Lauffer in Comprehensive Virology 17:1–82, 1981, with permission of the copyright owner, Plenum Publishing Corp.

($\overline{Q/L}$). This is done by measuring $\Delta S/\Delta t$ for a material of known diffusion coefficient, D, at a known value of Δc. After this apparatus constant is evaluated, D for different solutes can be evaluated.

DIFFUSION IN GELS

Because of the intricate tubular network in the living cell, the fluid content resembles a gel more than a simple solution or suspension. For that reason, it is necessary to consider the modifications of Fick's laws appropriate for diffusion in or into gels. This subject has been treated in detail by Lauffer [4] and by Schantz and Lauffer [5].

Consider first two solutions with solute concentrations c_a and c_b separated by a gel, as illustrated in Figure 3.3. A uniform concentration gradient, dc/dx, will be established in the steady state within the gel, equal in this case to $\Delta c/\Delta x$. Transport across this gel can be interpreted by Fick's first law, equation 3.3, except for the fact that diffusion is retarded by the gel, as was discussed in Chapter 2. Equation 2.29 shows how the frictional resistance to transport is increased by the presence of obstructing particles. Thus the rate of transport will be reduced. Equation 3.3 must be modified by multiplying by the reciprocal of F'/F given by equation 2.29

$$d\,\mathbf{S} = -D[(1 - \Phi)/(1 + (\alpha - 1)\,\Phi)]\;Q\;(dc/dx)dt$$

There is another problem. The concentration per unit volume of gel,

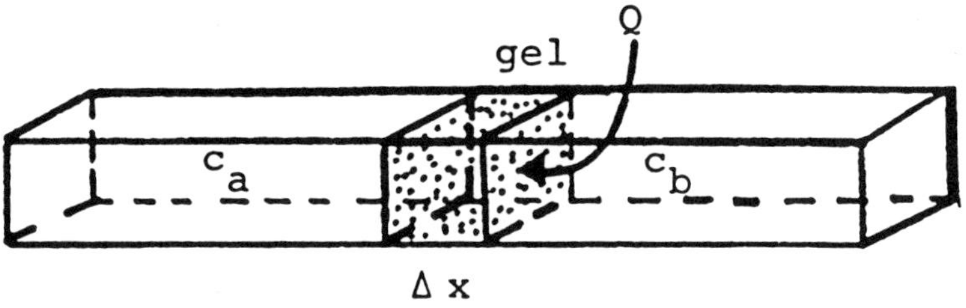

Fig. 3.3. Diagram to illustrate diffusion through a plug of gel from the left, where the concentration is c_a to the right, where the concentration is c_b. The concentration gradient in the liquid in the gel is $(c_b - c_a)/\Delta x$, a negative quantity. When concentration is expressed in terms of mass per unit volume of gel, the gradient is $(c_b - c_a)(1 - \Phi)/\Delta x$.

c′, is less than the concentration per unit volume of solution in the gel, c.

$c' = c(1 - \Phi)$ and $dc'/dx = dc/dx\ (1 - \Phi)$.
Thus, $d\,\mathbf{S} = -D[1/(1 + (\alpha - 1)\Phi]\ Q\ (dc'/dx)dt$.
When D′ is defined as $D[1/(1 + (\alpha - 1)\Phi)]$

$$d\,\mathbf{S} = -D'Q(dc'/dx)dt \tag{3.3b}$$

In these equations, Φ is the volume fraction of the gel substance, usually of the gel fibers. It is really the volume fraction excluded to the solute by the gel fibers.

Thus the concept of excluded volume, first encountered in the discussion of osmotic pressure in Chapter 1, has relevance to transport in gels. Since the gel material can be considered to be fibrous, the value of α can be taken to be that for randomly oriented rods, reported in Chapter 2 to be 5/3.

It was shown [4] by an argument like that used to derive equation 3.6 from equation 3.3 that Fick's second law for diffusion in a gel is

$$dc'/dt = D'\nabla^2 c' \tag{3.6a}$$

Thus Fick's first and second laws for diffusion in a gel are of exactly the same forms as for diffusion in a solvent. The advantage of this symmetry is that every equation derived from equations 3.3 and 3.6 can be applied to diffusion in gels by merely substituting c′ for c and D′ for D. If c′ had not been substituted for c in equation 3.3b, this symmetry would not have been obtained [4].[1]

SOLUTIONS OF FICK'S SECOND LAW

General solutions of equations 3.5 or 3.6 are beyond the scope of this book. However, solutions applicable to some problems can be arrived at by mathematical procedures involving relatively simple calculus. To pursue the simplest application, suppose that **S** moles of solute are deposited at t = 0 in a plane at x = 0 in a cylinder of unit cross section and infinite length, extending from $x = -\infty$ to $x = +\infty$. For this simple problem, equation 3.5 takes the form

$$\frac{\partial c}{dt} = D\,\frac{\partial^2 c}{\partial x^2}$$

It can be integrated[2] to yield

$$c = (A/\sqrt{t})e^{-x^2/4Dt}$$

when the diffusion coefficient, D, is constant and where A is an arbitrary constant of integration.

The arbitrary constant A can be evaluated as follows:

$$\mathbf{S} = \int_{-\infty}^{\infty} c\,dx = (A/\sqrt{t})\int_{-\infty}^{\infty} e^{-x^2/4Dt}\,dx$$

To evaluate the integral, let $u = x/\sqrt{4Dt}$; $dx = \sqrt{4Dt}\,du$

$$\mathbf{S} = 2A\sqrt{D}\int_{-\infty}^{\infty} e^{-u^2}\,du$$

The definite integral $\int_{-\infty}^{\infty} e^{-u^2}\,du$, frequently encountered in the mathematics of physical processes, has a value[3] of $\sqrt{\pi}$.

When this is substituted, A is found to have a value of $S/2\sqrt{\pi D}$ and

$$c = (S/2\sqrt{\pi Dt})e^{-x^2/4Dt} \tag{3.7}$$

Equation 3.7 shows how the **S** moles of solute will be distributed at the end of time t. The maximum concentration will occur at $x = 0$, the original location of all of it, and zero amount will occur at $x = \pm\infty$. Indeed, only very small amounts will occur at values of $\pm x$ for which $x^2/4Dt$ has a value of 5 or more.

There is another relatively simple way of solving equation 3.5 restricted to diffusion in the x-direction when the diffusion coefficient, D, is constant, a solution originally proposed by Boltzmann and more recently published by Longsworth [6]. It is clear from equation 3.7 that concentration varies with $(x/\sqrt{t})^2$ and is therefore a function of $x/\sqrt{t}$.

If $u \equiv x/\sqrt{t}$, c is a function of u.

$$\partial c/\partial t = (\partial u/\partial t)(\partial c/\partial u) = -(1/2)(x/t^{3/2})(\partial c/\partial u)$$

$$\partial c/\partial x = (\partial u/\partial x)(\partial c/\partial u) = (1/t^{1/2})(\partial c/\partial u)$$

$$\partial^2 c/\partial x^2 = (\partial/\partial x)(\partial c/\partial x) = (\partial u/\partial x)(\partial/\partial u)(\partial c/\partial x) = (1/t)(\partial^2 c/\partial u^2)$$

When these are substituted into equation 3.5 involving x only,

$$-(1/2)u\ \partial c/\partial u = D\ \partial^2 c/\partial u^2. \quad \text{Let } P \equiv \partial c/\partial u$$

$$-(1/2)\ uP = D\ \partial P/\partial u\ ;\ -(u/2)\ \partial u = D\ \partial P/P\ ;\ -(u^2/4) = D(\ln P - \ln A)$$

A is an arbitrary constant of integration

$$P = A\ e^{-u^2/4D} = \partial c/\partial u. \quad \text{Substitute definition of u.}$$

$$\partial c/\partial x = (A/\sqrt{t})\ e^{-x^2/4Dt} \tag{3.8}$$

The arbitrary constant A can be evaluated for various kinds of situations. A frequently used method of measuring diffusion coefficients is to establish a sharp boundary in a tube at $x = 0$ for $t = 0$ between a solution with solute concentration c_0 and solvent. During the diffusion process, concentration will be c_0 at $x = -\infty$ and 0 at $x = +\infty$. Now substitute $v \equiv x/\sqrt{4Dt}$ into equation 3.8.

$$\partial v/\partial x = 1/\sqrt{4DT} \text{ and } \partial c/\partial x = (\partial c/\partial v)(\partial v/\partial x)$$

$$(1/\sqrt{4Dt})\ \partial c/\partial v = (A/\sqrt{t})\ e^{-v^2} \quad ; \quad \int_{x=-\infty}^{x=\infty} dc$$

$$= A\sqrt{4D}\int_{-\infty}^{\infty} e^{-v^2}\, dv \quad ; \quad -c_0 = A\sqrt{4\pi D}$$

$$\partial c/\partial x = -(c_0/\sqrt{4\pi Dt})\ e^{-x^2/4Dt} \tag{3.9}$$

There are optical methods that can be used to measure dc/dx as a function of x after a time, t, of diffusion. Thus D can be evaluated.

The arbitrary constant A in equation 3.8 can be evaluated in a similar manner for other physical situations. Suppose, in one case, that some solute is produced in a steady state so that its concentration at a flat interface is always a constant, c_0, and that it then diffuses away into a solvent in which the concentration is 0 when $x = \infty$. Or, in another case, suppose that in a convection-free system where the concentration of a solute is c_0 at $x = -\infty$, solute is digested or precipitated or otherwise consumed at a flat interface so that $c = 0$ at $x = 0$. For both of these cases, A can be evaluated in the above manner to be $-2\ c_0/\sqrt{4\pi D}$. Thus,

$$\partial c/\partial x = -(2\ c_0/\sqrt{4\pi Dt})e^{-x^2/4Dt} \tag{3.10}$$

When it is desired to know concentration as a function of x, it is necessary to integrate equation 3.9 or 3.10 once more. One way of doing this is to define a new variable, z, not to be confused with the z axis of coordinates.

$$z \equiv x/\sqrt{2Dt} \tag{3.11}$$

TABLE 3.1. Abbreviated Numerical Solution of the Probability Integral

z	$\frac{1}{\sqrt{2\pi}}\int_z^\infty e^{-z^2/2}\,dz$	z	$\frac{1}{\sqrt{2\pi}}\int_z^\infty e^{-z^2/2}\,dz$
0.0	.5000	1.8	.0359
0.2	.4207	2.0	.0227
0.4	.3446	2.2	.0139
0.6	.2742	2.4	.0082
0.8	.2119	2.6	.0047
1.0	.1587	2.8	.0026
1.2	.1151	3.0	.0013
1.4	.0808	3.5	.0002
1.6	.0548	4.0	.0000

When equation 3.11 is substituted into equation 3.9,

$$dc = -(c_0/\sqrt{2\pi})\, e^{-z^2/2}\, dz.$$

When this is integrated between z where the concentration is c and ∞ where the concentration is 0,

$$c/c_0 = (1/\sqrt{2\pi}) \int_z^\infty e^{-z^2/2}\, dz \tag{3.12}$$

The integral on the right cannot be expressed in terms of a simple equation. It is the integral familiar to students of statistics, and tables are widely available giving the value of the integral for various values of z. When D and t are known, z can be calculated for every value of x by equation 3.11. Then from the table, the value of the integral can be read for the calculated value of z corresponding to a particular value of x. Then by equation 3.12 c/c_0 can be calculated for that value of x. A much abbreviated numerical solution of the integral in equation 3.12, the probability integral, is presented in Table 3.1.

A similar method, called the error function method, involves substituting $y = x/\sqrt{4Dt}$ into equation 3.9 to yield an equation similar to but not identical to equation 3.12. The determination of c/c_0 is similar to that just described except that error function, erf, tables or error function complement, erfc, tables must be used, depending on exactly how the problem is formulated. Care must be exerted not to use the wrong table!

An interesting application of equation 3.10 involves the determination of the total amount of solute, ΔS, that will diffuse in time, t, across a flat surface into a gel from a large volume of a well-stirred solution at concentration, c_0. The area of contact between the solution and the gel is Q and the boundary is at $x = 0$. For this problem, c must be replaced by c' and D by D'. At $x = 0$, equation 3.10 becomes $\partial c'/\partial x = -2c_0'/\sqrt{4\pi D't}$. Combine this with Fick's first law, equation 3.3b

$$d\,S = (D'Q\,2c_0'/\sqrt{4\pi D't})dt$$

$$\Delta\,S = (2Qc_0'/\sqrt{\pi})\sqrt{D't} \tag{3.13}$$

When the definitions of c' and D' used in the development of equation 3.3b are inserted, algebraic simplification is made and approximate roots are found.[4]

$$\Delta\,S \simeq 2(Qc_0/\sqrt{\pi})(1 - (4/3)\Phi)\sqrt{Dt} \tag{3.14}$$

This equation affords a simple method of measuring D. If a solution is placed above a thin gel in a Petri dish, convection will keep the concentration at c_0 immediately above gel surface. S is then measured by some appropriate analytical procedure after time, t. The time must be short enough to avoid appreciable reflection of solute from the bottom of the gel back into the solution [4]. One way of estimating Φ is to allow diffusion to take place long enough for complete equilibrium to be established. Then determine c, the amount per unit volume of solution and c', the amount per unit volume of gel, analytically. From the definition of c', $\Phi = 1 - c'/c$. This automatically takes care of the excluded volume problem.

RADIALLY SYMMETRICAL SPHERICAL DIFFUSION

Diffusion across a spherical boundary is more apt to be encountered in biological systems than diffusion across a plane interface. When the diffusion is entirely in the direction of radii of a sphere, that is, when it is radially symmetrical, relatively simple methods can be used to integrate Fick's law. For radially symmetrical diffusion, Fick's first law takes the form,

$$d\,S = -DQ(dc/dr)\,dt \tag{3.15}$$

By a method analogous to the derivation[5] of equation 3.5 from 3.3, Fick's second law takes the form

$$\partial c/\partial t = D\left(\frac{\partial^2 c}{\partial r^2} + \frac{2}{r}\frac{\partial c}{\partial r}\right) = D\nabla^2 c \tag{3.16}$$

By making the substitution, $\omega \equiv (c - c_0)\, r$, where c_0 is constant, equation 3.16 can be transformed[6] into

$$\partial\omega/\partial t = D\partial^2_\omega/\partial r^2 \tag{3.17}$$

Equation 3.17 is of the same form as equation 3.5 for diffusion in the x-direction only. Thus, by operations exactly like those used to obtain equation 3.8, one can derive,

$$\partial\omega/\partial r = (A/\sqrt{t})\, e^{-r^2/4Dt} \tag{3.18}$$

Now consider diffusion into a spherical particle of radius, a, when $c = c_1$, a constant, at $r = a$ and $c = c_0$ at r much greater than a.

$$\int_{r=a}^{r=\infty} d\omega = (A/\sqrt{t}) \int_{r=a}^{r=\infty} e^{-r^2/4Dt}\, dr = (A/\sqrt{e}) \int_{r-a=0}^{r-a=\infty} e^{-(r-a)^2/4Dt}\, d(r-a)$$

Substitute $u \equiv (r - a)/\sqrt{4Dt}$ and $\sqrt{4Dt}\, du = d(r - a)$

$$\int_{r=a}^{r=\infty} d\omega = A\sqrt{4D} \int_0^{\infty} e^{-u^2}\, du \quad ; \quad \omega \Big|_{r=a}^{r=\infty} = A\sqrt{\pi D}$$

Since $\omega = r(c - c_0)$ and since $c = c_0$ for all very large values of r, not only at $r = \infty$, $\omega = 0$ at $r = \infty$. At $r = a$, $\omega = a(c_1 - c_0)$ Therefore,

$$A = -(c_1 - c_0)\, a/\sqrt{\pi D}$$

$$\partial\omega/\partial r = -[(c_1 - c_0)\, a/\sqrt{\pi D t}]\, e^{-r^2/4Dt} \text{ or } \partial\omega/\partial(r - a)$$

$$= -[(c_1 - c_0)\, a/\sqrt{\pi D t}]\, e^{-(r-a)^2/4Dt} \tag{3.19}$$

Suppose the sphere, bathed by a solution with solute concentration c_0, instantly absorbs or combines with or digests the solute so that the concentration in solution at $r = a$ is always zero. c_1 is 0 and c at $r = a$ is 0. From the definition of ω and equation 3.19,

$$\partial\omega/\partial r = (c - c_0) + r\frac{\partial c}{\partial r} = -[((c_1 - c_0)a)/\sqrt{\pi Dt}]\, e^{-(r - a)^2/4Dt}$$

When $r = a$

$$-c_0 + a\,\partial c/\partial r = c_0\, a/\sqrt{\pi Dt}$$

$$(\partial c/\partial r)_a = c_0\,(1/a + 1/\sqrt{\pi Dt}) \qquad (3.20)$$

To reach the surface, solute must pass through the gradient $(\partial c/\partial r)_a$. From equation 3.15 with sign changed because c decreases as r decreases,

$$d\,\mathbf{S} = DQ\,(\partial c/\partial r)\,dt = 4\pi a^2\,Dc_0\,[1/a + (1/\sqrt{\pi Dt})]\,dt \qquad (3.21)$$

Since $\mathbf{S} = 0$ when $t = 0$,

$$\Delta\,\mathbf{S} = 4\pi a\,Dc_0\left(t + \frac{2a\sqrt{t}}{\sqrt{\pi D}}\right) \qquad (3.22)$$

$\Delta\,\mathbf{S}$ is the number of moles of solute "destroyed" is time, t. This equation was derived in a different manner by v. Smoluchowski [7].

In a similar manner it can be shown that when c_0 at large values of r is 0 and concentration at $r = a$ is maintained constant at c_1, $-(\partial c/\partial r)_a = c_1\,(1/a + 1/\sqrt{\pi Dt})$. This equation would apply to a situation in which a solute is synthesized at a constant rate inside the sphere and diffuses away into the solvent.

The relevance of equation 3.22 to biology can be demonstrated by considering the following hypothetical, but realistic, problem involving the adsorption of bacteriophages to susceptible bacteria. Suppose that n bacteriophages with $D = 3 \times 10^{-8}$ cm^2/ sec are mixed with m susceptible spherical bacteria of radius, a, equal to 10^{-4} cm, in 1 ml. Whenever a phage particle is adsorbed by a bacterium, it is destroyed in the sense that it is removed from solution. For one bacterium, $\Delta \mathbf{S} = \boldsymbol{k}\, \Delta n$ and for all bacteria, $\Delta N = m\boldsymbol{k}\, \Delta n$. c_0 must be replaced by $\boldsymbol{k}$ n, where $\boldsymbol{k}$ is a constant of proportionality. Thus, from equation 3.22

$$\Delta N = m\, 4\pi a\, Dn(t + 2a\sqrt{t}/\sqrt{\pi D})$$

If m is 10^{10} bacteria per ml, how long will it take for half of the bacteriophages to be adsorbed? This corresponds to the case that $\Delta N/n = 1/2$. When the values of a, D and m are substituted, $(t + 0.65\sqrt{t}) = 10^{12}/75.4 \times 10^{10} = 1.33$. By trial and error, it can be solved for t to yield a value of $t = 0.76$ sec. This number should be reassuring to virologists.

A second example of biological relevance involves the concept, frequently encountered in biological literature, of diffusion-limited reaction rates. To develop this concept, one starts with equation 3.21. This equation contains the term $(1/a + 1/\sqrt{\pi DT})$. Spherical particles as large as $a = 1.4 \times 10^{-7}$ cm, corresponding to a molecular weight of about 10^4 daltons, have a diffusion coefficient, calculated from the Einstein-Sutherland equation, equation 3.4, and Stokes' law, equations 2.15 and 2.16, of about 1.5×10^{-6} cm^2/sec. If $t = 10^{-3}$ sec, $1/\sqrt{\pi Dt}$ is 0.15×10^5. But $1/a = 1/1.4 \times 10^{-7} \simeq 70 \times 10^5$. Thus, even for particles as large as 1.4×10^{-7} cm in radius and for times as short as 10^{-3} sec, $1/a$ is much larger than $1/\sqrt{\pi Dt}$. Thus the differential equation 3.21 can be simplified for small spheres to

$$d\,\mathbf{S} = 4\pi a\, Dc_o dt$$

When identical spheres collide, the sphere of closest approach has a radius of 2a, an example of the excluded volume principle introduced in Chapter 1. Also, the number of moles disappearing as a result of collisions, $-d\mathbf{S}'$, is 2 $d\mathbf{S}$, because both molecules involved are changed by reaction and thus disappear. Thus, $-d\mathbf{S}' = 16\, \pi a D c_0 dt$ per molecule. The total number of molecules is $c_0 \mathbf{A} V$, and the total number of moles disappearing, $-d\mathbf{S}''$ is $-d\mathbf{S}'' = 16\, \pi a D c_0^2\, \mathbf{A}\, V dt$. The reaction velocity, $-dc/dt = -(d\mathbf{S}''/dt)(1/V)$ and $c = m/1{,}000$, where m is the molarity.

$$-dm/dt = 16\pi aD\,\mathbf{A}\,m^2/1{,}000 \equiv k\,m^2$$

where k is the second order reaction velocity constant with dimensions of molarity^{-1} time^{-1}.

$$k = 16\,\pi aD\,\mathbf{A}/1000.$$

From equations 3.4, 2.15, and 2.16, $D = RT/6\pi a\eta\mathbf{A}$.

$$k = 16\,RT/6{,}000\,\eta = (8/3) \times 10^{-3}\,RT/\eta \qquad (3.23)$$

Thus, it is noteworthy that the diffusion-limited rate constant is independent of the size of the spheres, provided only that they are small. When the values of R, T, and η for water at 20°C are inserted, k has a value of 0.65×10^{10}. Most bimolecular reactions encountered in biological systems are slower than this and are, therefore, not diffusion-limited. This means that reaction does not follow every collision.

RADIALLY SYMMETRICAL CYLINDRICAL DIFFUSION

For radially symmetrical diffusion across a cylindrical surface, Fick's first law is expressed by equation 3.15. By a method analogous to the derivation[7] of equation 3.5 from equation 3.3 and of equation 3.16 from equation 3.15, Fick's second law for this can take the form,

$$\frac{\partial c}{\partial t} = (D/r)\,\frac{\partial}{\partial r}\left(r\frac{\partial c}{\partial r}\right) = D\left(\frac{\partial^2 c}{\partial r^2} + \frac{1}{r}\frac{\partial c}{\partial r}\right) \equiv D\nabla^2 c \qquad (3.24)$$

Most of the useful solutions of equation 3.24 involve Bessel functions, not really simple calculus and thus beyond the scope of this book. The reader can find numerous examples in Chapter V of *The Mathematics of Diffusion,* by Crank [8].

Because cylindrical structures like muscle fibers and nerve cells are so frequently encountered in organisms, it would be improper to dismiss this subject entirely merely because of mathematical complex-

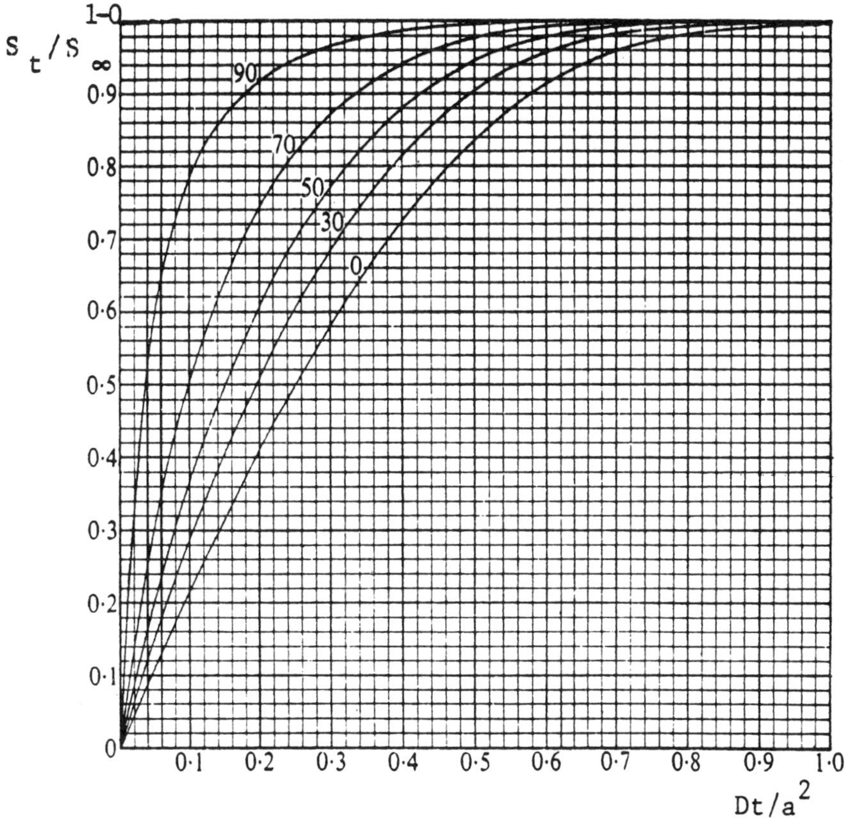

Fig. 3.4. Uptake by a cylinder from a stirred solution of limited volume. Numbers on curves show percentages of total solute finally taken up by cylinder. Reprinted from Crank [8], with the permission of the copyright owner, Oxford Press.

ity. Crank [8] has reproduced graphical solutions for several physical situations, two of which have potential relevance to biological systems. The first of these is shown in Figure 3.4. It corresponds to the situation in which a cylinder of radius, a, initially free of solute, is immersed in a larger concentric cylinder containing solute at an initial concentration of c_0. The cross-sectional area between the concentric cylinders, the region containing the solute initially at c_0, is Q. The external solution is constantly stirred and the inner cylinder is convection-free. $\mathbf{S}_t$ is the amount of solute taken up by the inner cylinder in time, t, and $\mathbf{S}_\infty$ is the amount taken up in infinite time. If there is a partition factor, κ, which will be unity for many systems, at final equilibrium the concentration inside will be κc_∞ and that outside will be c_∞. Since mass is conserved, $\pi a^2 L c_\infty \kappa + Q L c_\infty = S_0$, where L is the length of the cylinder and S_0 is the total amount of solute.

$$\pi a^2 L c_\infty \kappa + \frac{Q}{\pi a^2 \kappa} \pi a^2 L c_\infty \kappa = S_0. \ \pi a^2 L c_\infty \kappa \text{ is } S_\infty \text{ and } \frac{Q}{\pi a^2 \kappa} \text{ is defined as } \alpha$$

$$\therefore \quad S_\infty + \alpha S_\infty = S_0$$

$$S_\infty / S_0 = 1/(1 + \alpha)$$

Figure 3.4 shows S_t/S_∞ as a function of the dimensionless variable, $\sqrt{Dt/a^2}$ for five values of S_∞ / S_0 indicated on the curves as percentages. The curve labeled 0 corresponds to the situation in which the external volume is so large that only a negligible fraction of the total solute is taken up by the inner cylinder. Since living cells have sufficient internal structure to minimize convection, this graph could describe the uptake of solute from a large volume of circulating fluid containing the solute at a constant concentration, c_0.

The same graphs describe the course of diffusion outward from the inner cylinder at an initial concentration into a well-stirred outer solution initially free of solute. The only difference is that S_∞ / S_0 is now $1/(1 + 1/\alpha)$.

A second potentially useful solution, corresponding to the situation in which the initial concentration throughout the region where r is greater than a is c_0 and the surface of the cylinder at $r = a$ is maintained at a constant concentration c_1, is shown in Figure 3.5. Here $(c - c_0)/(c_1 - c_0)$ is shown as a function of r/a for values top to bottom of 4, 1 and ¼ of Dt/a^2. It would correspond to the case in which a metabolite is being produced at a constant rate inside the cylinder and is diffusing outward into a convection-free external solution. It should also correspond to the case where solute is diffusing into the cylinder, there to be

Fig. 3.5. Concentration-distance curves in the region $r > a$. Numbers on curves are values of Dt/a^2, 4 on top, 1 in the middle and ¼ on the bottom. Reprinted from Crank [8], with the permission of the copyright owner, Oxford Press.

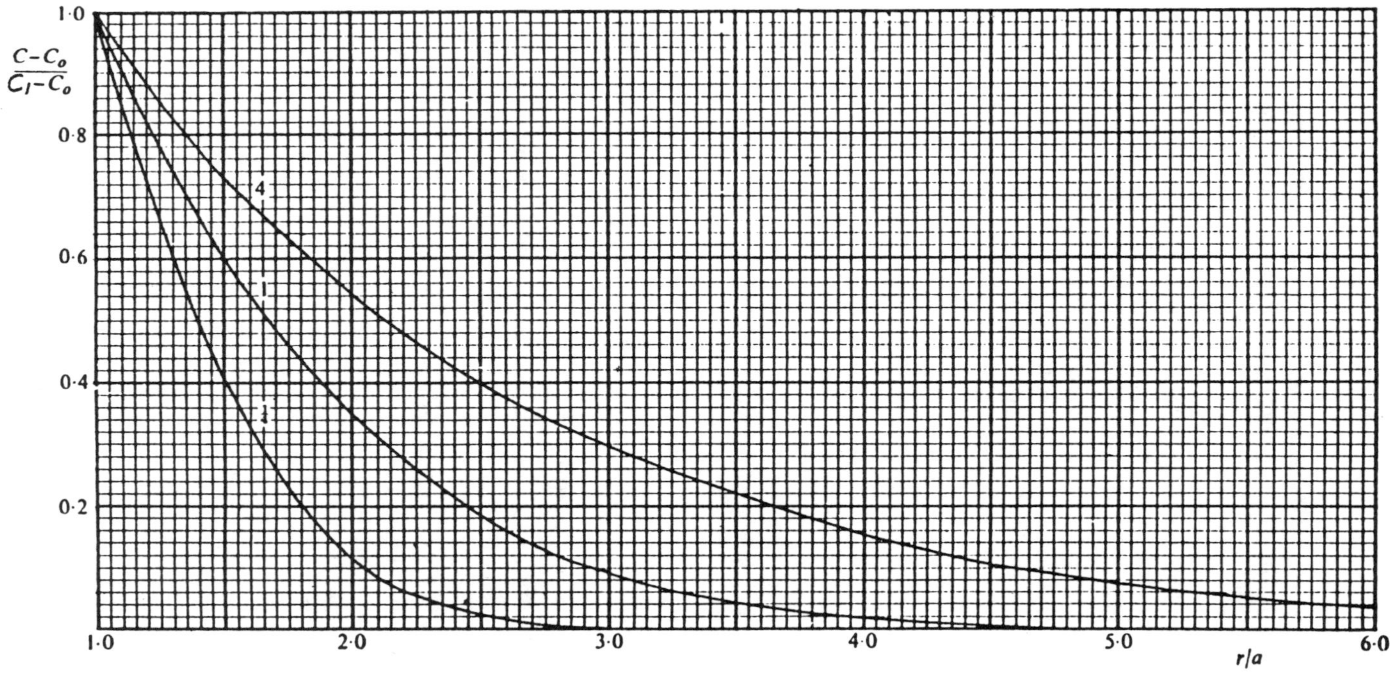
1·0
0·8
0·6
0·4
0·2
$\frac{C-C_o}{C_I-C_o}$
4
1
1·0
2·0
3·0
4·0
5·0
6·0
r/a

consumed so as to maintain the surface concentration, c_1 constant, possibly 0. This case is comparable to that represented by equation 3.12 for a flat interface.

A few simple solutions for cylindrical diffusion are available. One of these involves diffusion across a cylindrical shell or membrane. In the steady state, $\partial c/\partial t = 0$, that is, the total amount of solute in the shell remains constant. By equation 3.24, $(\partial/\partial r)r(\partial c/\partial r)$ is therefore 0. The general solution is $c = A + B \ln r$, where A and B are arbitrary constants. If the concentration is kept at c_1 at the inner surface, $r = a$, and at c_2 at the outer surface, $r = b$, the equation becomes[8]

$$c = [c_1 \ln (b/r) + c_2 \ln (r/a)]/\ln (b/a) \qquad (3.25)$$

for values of r between a and b.

$$dc/dr = (c_2 - c_1)/r \ln (b/a).$$

When this and $Q = 2\pi rL$ are substituted into equation 3.15 with sign changed because dr is negative,

$$dS = 2\pi rLD(c_2 - c_1)\, dt/r \ln (b/a) \text{ and}$$

$$\Delta S = 2\pi LD\Delta t\, (c_2 - c_1)/\ln (b/a) \qquad (3.26)$$

Equation 3.26 would be valid for the diffusion of a substance through the membrane of a nerve cell if the membrane were permeable to that substance, if the cell were bathed on the outside by a solution at constant concentration, c_2 and if it metabolized the substance on the inside to keep c_1 a constant, perhaps nearly zero. If $(b - a)/a$ is much less than 1, $\ln (b/a) = \ln (1 + (b - a)/a) \hat{=} (b - a)/a$. Under these conditions, equation 3.26 becomes practically the same as that for diffusion across a flat membrane. D in any case must be the diffusion coefficient in the membrane.

An approximate equation, reasonably accurate for very thin cylinders, can be derived by a method similar to that used in the derivation of equation 3.7 for the diffusion of S moles deposited in a plane after diffusion in one direction for time t. This time, start with M moles deposited in a point in an infinite plane, x, y. For this physical situation, equation 3.5 takes the form

$$\partial c/\partial t = D(\partial^2 c/\partial x^2 + \partial^2 c/\partial y^2)$$

The solution[9] is $c = (A/t)e^{-(x^2+y^2)/4Dt}$
The arbitrary constant, A, equals[9] $\mathbf{S}/4\pi D$. Thus, when $x^2 + y^2 = r^2$

$$c = (\mathbf{S}/(4\pi Dt))e^{-r^2/4Dt} \qquad (3.27)$$

For this physical situation, c is defined as moles per unit area. This same equation applies to radially symmetrical diffusion when **S** moles of solute are deposited per unit length of a line and diffusion takes place into an infinite volume. The reason is that **S** can be considered to be the sum of an infinite number of small sources. The only gradients will be in the r direction—all points at a fixed value of r in the direction of the line will be at the same concentration. The sum of the cs per unit area along unit length will be the concentration per unit volume, c as normally defined.

If, instead of depositing **S** moles on a line, one deposits them in a thin cylinder, equation 3.27 will approximate the values of c at different values of r after time, t, of diffusion. The approximation increases in accuracy as the ratio of r to the radius of the cylinder increases.

CONCENTRATION-DEPENDENT DIFFUSION

All of the equations derived thus far have been developments from equation 3.3 in which the coefficient of diffusion, D, is a constant at constant temperature and in which $d\Pi/dx = RTdc/dx$. The many derived equations are strictly valid only when these conditions obtain. However, the most fundamental starting point is equation 3.1 expressed in terms of $d\Pi/dx$ instead of dc/dx. It is clear from equation 1.10 as rewritten in the third paragraph of this chapter that Π must be represented as a power series in c and Φ. Since Φ is proportional to c, it can be reduced to a power series in c, and $d\Pi/dx$ would also be a power series in c. The f in equation 3.1 really means the actual coefficient of friction that can be represented by $f' \cdot f' = f$, the coefficient of friction at infinite dilution, multiplied by (F'/F), given by equation 2.29. This is true because at even moderate concentrations the diffusing particles are moving in an obstructed system. f' can also be expressed as f, a constant, multiplied by a power series in c. Since one of these power series is in the numerator and the other in the denominator, they cancel each other partially but not completely, since they are not identical. The net result is that even in moderately concentrated solutions, equation 3.3 represents a valid approximation in many cases. The mathematics of dealing with equation 3.3 modified by the inclusion of a power series in c is beyond the scope of our present treatment.

There is, however, one circumstance under which the deviations from equation 3.3 cannot be considered to be trivial. As was illustrated in Figure 1.3, the deviation from the simple principle that Π is proportional to c can become enormous for charged macromolecules in dilute salt solutions. This is directly related to the Donnan term, the final one on the right in equation 1.10. The diffusion of charged macromolecules requires further discussion, as presented in the next section.

DIFFUSION OF CHARGED MACROMOLECULES

While consideration of the diffusion of charged macromolecules can be treated in terms of equation 1.10, a more intuitive analysis might serve better to illustrate the physical principles involved. Imagine a simple salt that ionizes to form two ions of equal valence, one small and the other large. The two ions will tend to diffuse separately, the small one more rapidly than the large one. Yet electrical neutrality must be preserved. The diffusion of the small ion will be retarded by the oppositely charged large ion and that of the large ion will be speeded up. The net diffusion rate of the salt will then be intermediate between the hypothetical diffusion rates of the fast and the slow ions. Many, if not most, of the macromolecules important in biology are highly charged and, therefore, except at the isoelectric point, must be considered to be multivalent salts of small cations or anions. The same principle must apply as in the case of the simple ion discussed above. The small counterions will tend to diffuse more rapidly than the oppositely charged macromolecules and will thus tend to increase the observed diffusion rate of the macroion.

One intuitive way of discussing this situation further is to recognize that the tendency of the counterions to diffuse more rapidly than the macroion will establish a potential gradient. As will be discussed more fully in the next chapter, a potential gradient acts on a charged particle to produce a force, a force that in this case would speed up the macroion. However, the magnitude of the potential gradient established by this tendency of the small counterions to move away from the macroion is inversely proportional to the conductivity of the solution. Per unit of cross sectional area, the number of charges on opposite ions that tend to diffuse through in unit time minus the smaller number of charges on the macromoles that tend to follow more slowly constitute an electric current, **I**. By Ohm's law, **V** = **IR** and d**V**/dx = **I**d**R**/dx. d**V**/dx is the potential gradient, and, for unit area of cross section, d**R**/dx is the specific resistance, the resistance per unit volume. This, in turn, is the reciprocal of the specific conductivity, **λ**. Hence, the potential gradient is inversely proportional to specific conductivity. For a given salt, **λ** is approximately proportional to concentration.

Lower potential gradients will be established between counterions and macroions in solutions of high conductivity than in solutions of low conductivity. Thus the accelerating force on the macroion will be much less in solutions of high than of low conductivity. The practical solution, then, when one attempts to measure the diffusion coefficient of a macroion, is to carry out the experiments in solutions of reasonably high conductivity. Experience has shown that measurements made on charged proteins in solutions of small electrolytes at concentrations of the order of 0.1 M yield results not significantly different from those obtained with isoelectric proteins. Thus the problem can be solved simply in well-designed experiments. This analysis is entirely consistent with the expectations based on equation 1.10. Reference to Figure 1.3 will show that for moderately concentrated macroions in 0.10 M electrolyte, the departure from the proportionality of Π to c is small. In real biological systems, however, one has to deal with the actual small electrolyte concentrations in the system. The osmotic pressure of vertebrate blood is approximately equivalent to that of a 0.1 M NaCl solution. The fluids in land plant tissues exhibit a considerable range of osmotic pressures, from half that of vertebrate blood to many times that figure. In vertebrate fluids, charged macromolecules having a charge of ± 25 proton units, as in the example illustrated in Figure 1.3, when present at concentrations below a few per cent, should diffuse at approximately the same rate as the same molecule in the isoelectric state. However, charged macromolecules with valence significantly greater than ± 25 could exhibit accelerated diffusion rates at concentrations higher than a few per cent.

THERMAL DIFFUSION

Mita et al. [9] raised the question whether active membrane transport could result from the flux of metabolic heat. They point out that only extremely small, practically unmeasurable, temperature differences between the inside and outside of a cell could result from metabolic activity within the cell. Nevertheless, because of the thinness of the cell membrane, the temperature gradient across the membrane could be significant. In their article, Mita et al. review evidence that there are changes in ion transport across biological membranes in response to artificially produced temperature gradients.

The early literature relevant to liquid thermal diffusion has been surveyed by Moseley [10]. He points out that if a temperature gradient is maintained in an originally homogeneous solution it is found that, after a time, a concentration gradient will also exist. The phenomenon was first observed by Ludwig, but it is usually referred to as the Soret effect after its second discoverer. The first theoretical explanation of the effect was offered by Van't Hoff in 1887. On the assumption that the osmotic pressure, Π, must be constant throughout the solution, and since Π equals cRT, at equilibrium the product of cT must be constant throughout the solution. At the lower of two temperatures, c must be higher than at the higher temperature. Some early experimental results agree with this theory reasonably well; others do not.

Katchalsky and Curran [11] discuss solute transport in systems with temperature gradients in the language of nonequilibrium thermodynamics. They arrive at the result that the flux of the solute, the amount flowing across unit cross section in unit time, is equal to minus the diffusion coefficient times the gradient of concentration minus the product of the thermal diffusion coefficient and the concentration times the gradient of the temperature. The first of these terms is Fick's law of diffusion, equation 3.3. The second term represents thermal diffusion.

The ratio of the thermal diffusion coefficient to the ordinary diffusion coefficient is called the Soret coefficient. Katchalsky and Curran go on to point out that thermal diffusion in binary systems has been studied extensively in liquid systems. A cell is filled with a solution of uniform composition and the two end plates are maintained at different temperatures. A steady state concentration gradient develops in the presence of the constant temperature gradient. The ratio of these gradients is equal to $-cD_T/D$ where D_T is the thermal diffusion coefficient. The thermal diffusion coefficient is normally found to be smaller by a factor of 10^2 to 10^3 than the ordinary diffusion coefficient. As will be shown in the next paragraph, this result is consistent with deductions from osmotic pressure theory. Such a deduction does have the virtue, not contained in the general treatment of the subject from the point of view of nonequilibrium thermodynamics, of providing a rational physical basis for thermal diffusion.

As was shown in the derivation of equation 3.1, when c is moles of solute per unit volume of solution, the force, F, exerted on a single solute molecule is

$$F = -(d\Pi/dx)/c\mathbf{A} \qquad (3.28)$$

As $c \rightarrow 0$, $\Pi = RTc$.

For a fixed value of c but where T varies with x,

$$d\Pi/dx = Rc\ dT/dx$$

$$\therefore \quad F = -Rc(dT/dx)/c\mathbf{A} = -\mathbf{k}(dT/dx) \qquad (3.29)$$

At the steady state, the frictional force, fv, is equal to F.
In time, dt, v dt Qc moles of solute $\equiv$ dS will pass through an area Q.
$dS = v\ Qc\ dt = (F/f)cQ\ dt = -(\mathbf{k}c/f)Q(dT/dx)dt$

$$dS = -D_T c\ Q(dT/dx)dt \qquad (3.30)$$

$D_T \equiv (\mathbf{k}/f)$ and is the thermal diffusion coefficient.
Π is osmotic pressure, **k** is the Boltzmann constant, f is the coefficient of friction, and **A** is Avogadro's number.

Since, by the Einstein-Sutherland relationship, equation 3.4, the ordinary diffusion coefficient, $D = \mathbf{k}T/f$, at biological temperature, $D \simeq 300\ D_T$, clearly in the range of 10^2–10^3 mentioned above. For any membrane of thickness Δx, equation 3.30 can be integrated to give $\Delta S = -D_T c_i\ Q\ (\Delta T/\Delta x)\ \Delta t$, where c_i is the concentration on the high temperature side. Fick's first law, equation 3.3, can be integrated to give $\Delta S = DQ\ (\Delta c/\Delta x)\ \Delta t$. For thermal diffusion to be of the same magnitude, but opposite in direction, $D_T c_i\ \Delta T$ must equal $D\Delta c$. Since $D/D_T = 300$, $\Delta c/c_i = \Delta T/300$. Only when ΔT is very large can $\Delta c/c_i$ be of appreciable magnitude at the steady state. This argues against typical thermal diffusion playing an appreciable role in biological transport.

REFERENCES

1. Fick, A. Ann. Phys. Leipzig *170*:59, 1855.
2. Fourier, J.B. Théorie Analytique de la Chaleur (Oeuvres de Fourier), 1822.
3. Northrop, J.H. and Anson, M.L. J. Gen. Physiol. *12*:543, 1929.
4. Lauffer, M.A. Biophys. J. *1*:205–213, 1961.
5. Schantz, E.J. and Lauffer, M.A. Biochemistry *1*:658–663, 1962.
6. Longsworth, L.G., Ann. NY Acad. Sci. *46*:211, 1945.
7. v. Smoluchowski, M. Z. Phys. Chem. *92*:129, 1917.
8. Crank, J. The Mathematics of Diffusion, London Oxford University Press, 1956.
9. Mita, D.G., Canciglia, P., D'Acunto, A., and Gaeta, F.S. Comments on Molecular and Cellular Biophysica *3*:195–217, 1986.
10. Moseley, H.M. In: Liquid Thermal Diffusion, P.H. Abelson, N. Rosen, and I. Hooper, (eds.). Washington Naval Research Laboratory, pp. 11ff, 1958.
11. Katchalsky, A. and Curran, P.F. Non-equilibrium Thermodynamics in in Biophysics, Cambridge, Harvard University Press, 1965.

NOTES

1. This point can be illustrated by substituting $(1 - \Phi)$ dc/dx for dc′/dx into equation 3.3b. It then takes the form $d\mathbf{S} = -D'(1 - \Phi)\,Q\,(dc/dx)dt$. However, when $c\,(1 - \Phi)$ is substituted for c′ in equation 3.6a, it becomes $dc/dt = D'\nabla^2 c$. Thus the diffusion coefficient in Fick's first law would be $D'(1 - \Phi)$, but in Fick's second law, simply D′. While this is mathematically correct, it is aesthetically preferable to formulate the problem in such a way that the coefficient is the same in Fick's first and second laws.

2. This can be proved by differentiation. When $c = (A/\sqrt{t})e^{-x^2/4Dt}$

$$\partial c/\partial t = -\frac{1}{2} A\,t^{-3/2}\,e^{-x^2/4Dt} + A\,t^{-1/2}\,e^{-x^2/4Dt}\,(x^2/4Dt^2)$$

$$= At^{-3/2}\,e^{-x^2/4Dt}\left(\frac{x^2}{4Dt} - \frac{1}{2}\right)$$

$$\partial c/\partial x = -(At^{-3/2}/2D)xe^{-x^2/4Dt}$$

$$\partial^2/\partial x^2 = -(At^{-3/2}/2D)(e^{-x^2/4Dt} - (2x^2/4Dt)\,e^{-x^2/4Dt})$$

$$\partial^2 c/\partial x^2 = (At^{-3/2}/D)\, e^{-x^2/4Dt}\left(\frac{x^2}{4Dt} - \frac{1}{2}\right)$$

$$\therefore \quad \partial c/\partial t = D\frac{\partial^2 c}{\partial x^2}$$

3. The definite integral,

$$\int_{-\infty}^{\infty} e^{-u^2}\, du$$

is, because of symmetry,

$$2\int_{0}^{\infty} e^{-u^2}\, du$$

Since functions of the same form when integrated between the same limits have the same constant value,

$$\int_{0}^{\infty} e^{-u^2}\, du = \int_{0}^{\infty} e^{-x^2}\, dx = \int_{0}^{\infty} e^{-y^2} dy$$

$$\left\{\int_{0}^{\infty} e^{-u^2}\, du\right\}^2 = \int_{0}^{\infty}\int_{0}^{\infty} e^{-(x^2+y^2)}\, dxdy$$

The double integral can be evaluated by a method attributed to Gauss. It represents an area within one quadrant. When expressed in polar coordinates, $x^2 + y^2 = r^2$ but the infinitesimal area element, dxdy, is replaced by $rd\theta dr$. The limits for the quadrant in polar coordinate are 0 to ∞ for r and 0 to $\pi/2$ for θ

$$\left\{\int_{0}^{\infty} e^{-u^2}\, du\right\}^2 = \int_{0}^{\pi/2}\int_{0}^{\infty} e^{-r^2}\, r\, dr\, d\theta = \frac{\pi}{2}\int_{0}^{\infty} e^{-r^2}\, r\, dr = \frac{\pi}{4}$$

$$\int_0^{\infty} e^{-u^2}\,du = \sqrt{\pi}/2 \text{ and } \int_{-\infty}^{\infty} e^{-u^2}\,du = \sqrt{\pi}$$

4.

$$\Delta S = (2Qc_0\,(1 - \Phi)/\sqrt{\pi})\,\sqrt{t}\,\sqrt{D/(1 + (\alpha - 1)\,\Phi}$$

For randomly oriented rods $\alpha = 5/3$

$$\Delta S = (2Qc_0/\sqrt{\pi})\,\sqrt{Dt}\left(\left(1 - \Phi\right)\Big/\sqrt{1 + \frac{2}{3}\Phi}\right)$$

$$\sqrt{1 + \frac{2}{3}\Phi} \simeq 1 + \frac{1}{3}\Phi \quad ; \quad (1 - \Phi)\Big/\left(1 + \frac{1}{3}\Phi\right) \simeq 1 - \frac{4}{3}\Phi$$

5. Imagine spherically symmetrical diffusion across the surface of a sphere at radius r into a spherical shell at radius r + dr.

$$\begin{aligned}\partial c &= [(dS)_r - (dS)_{r+dr}]/V = (dS)_r - [(dS)_r + (\partial/\partial r)(dS)dr]/v \\ &= -\,(\partial/\partial r)(dS)dr/V\end{aligned}$$

Since $Q = 4\pi r^2$ and V, the volume of the shell, $= 4\pi r^2 dr$

$$\partial c = -\,\partial/\partial r\,[-4\pi D r^2\,(\partial c/\partial r)dt]\,dr/4\pi r^2\,dr = (D/r^2)\,\partial/\partial r\,[r^2\,(\partial c/\partial r)]dt$$

$$\partial c/\partial t = D\,[\partial^2 c/\partial r^2 + (2/r)(\partial c/\partial r)] \equiv D\nabla^2 c$$

6.

$$\omega \equiv (c - c_o)r \quad ; \quad \partial\omega/\partial t = r(\partial c/\partial t) \quad ; \quad \partial c/\partial t = (1/r)(\partial\omega/\partial t)$$

$$\partial\omega/\partial r = (c - c_o) + r\,(\partial c/\partial r) \quad ; \quad \partial^2\omega/\partial r^2 = r(\partial^2 c/\partial r^2) + 2\,(\partial c/\partial r)$$

When these are substituted into equation 3.16,

$$\partial\omega/\partial t = D\partial^2\omega/\partial r^2$$

7. Imagine radially symmetrical diffusion across the surface of a cylinder of length L at radius r into a cylindrical shell at radius r + dr.

$$\partial c = [(dS)_r - (dS)_{r+dr}]/V = (dS)_r - [(dS)_r + (\partial/\partial r)(dS)\,dr]/V = -\left(\frac{\partial}{\partial r}\right)(dS)\,dr/V$$

Since $Q = 2\pi rL$ and $V = 2\pi rLdr$

$$\partial c = -(\partial/\partial r)[-2\pi LDr(\partial c/\partial r)]\,\partial t\,dr/2\pi rLdr = (D/r)\,(\partial/\partial r)\left(r\frac{\partial c}{\partial r}\right)dt$$

$$\partial c/\partial t = (D/r)(\partial/\partial r)(r\,\partial c/\partial r) = D(\partial^2 c/\partial r^2 + (1/r)(\partial c/\partial r)) \equiv \nabla^2 c$$

8. That $c = A + B \ln r$ is a solution of $\partial/\partial r\,(r\,\partial c/dr) = 0$ is proved by differentiation. $\partial c/\partial r = B/r$, $r\,\partial c/\partial r = B$, and $\partial/\partial r\; r\,\partial c/\partial r = 0$. When the boundary conditions are substituted into $c = A + B \ln r$,

$$c_1 = A + B \ln a \text{ and } c_2 = A + B \ln b$$

$$(c_2 - c_1) = B \ln (b/a) \text{ and } B = (c_2 - c_1)/\ln (b/a)$$

When the value of B is now substituted into the general solution,

$$c_1 = A + (c_2 - c_1) \ln a /\ln (b/a) \text{ and } A = [c_1 \ln b/a - (c_2 - c_1) \ln a]/\ln (b/a)$$

Now substitute both A and B into $c = A + B \ln r$ to obtain

$$c = [c_1 \ln b/a - (c_2 - c_1) \ln a + (c_2 - c_1) \ln r]/\ln (b/a = [c_1 \ln (b/r) +$$

$$c_2 \ln (r/a)]/\ln (b/a)$$

9. When $c = (A/t)e^{-(x^2+x^2)/4Dt}$

$$\partial c/\partial t = A\, e^{-(x^2+y^2)/4Dt}\, [(x^2 + y^2)/4Dt^3 - 1/t^2]$$

$$\partial c/\partial x = \frac{-2Ax}{4Dt^2}\, e^{-(x^2+y^2)/4Dt}$$

$$\partial^2 c/\partial x^2 = A\, e^{-(x^2+y^2)/4Dt} \left(\frac{x^2}{4D^2t^3} - \frac{1}{2Dt^2}\right)$$

In an identical manner, it follows that

$$\partial^2 c/\partial y^2 = A\, e^{-(x^2+y^2)/4Dt} \left(\frac{y^2}{4Dt^3} - \frac{1}{2Dt^2}\right)$$

$$D(\partial^2 c/\partial x^2 + \partial^2 c/\partial y^2) = A\, e^{-(x^2+y^2)/4Dt} \left(\frac{x^2 + y^2}{4Dt^3} - \frac{1}{t^2}\right) = \partial c/\partial t \text{ Q.E.D.}$$

To evaluate the arbitrary constant A, note that

$$S = \int_{-\infty}^{\infty}\int_{-\infty}^{\infty} c\, dxdy = (A/t) \int_{-\infty}^{\infty}\int_{-\infty}^{\infty} e^{-(x^2+y^2)/4Dt}\, dxdy.$$

$$S = (A/t)\, [\sqrt{4Dt} \int_{-\infty}^{\infty} e^{-x^2/4Dt}\, dx/\sqrt{4Dt}]\, [\sqrt{4Dt} \int_{-\infty}^{\infty} e^{-y^2/4Dt}\, dy/\sqrt{4Dt}]$$

As was seen in the derivation of equation 3.7, each of these definite integrals equals $\sqrt{\pi}$. Thus, $A = S/4\pi D$.

4

Motion in Electric Fields

Electric fields are widespread in biological systems, especially near cell membranes. Biological macromolecules are frequently charged. Capillaries are lined with fixed electric charges. Ordinary electrolytes ionize. There are widespread opportunities for the movement of fluids, ions, and macroions in electric fields. This chapter will deal with simple electrolyte solutions. The following chapter will present a discussion of electrophoresis, the movement of charged macroions or larger particles, and electro-osmosis, the movement of fluids through charged capillaries, both in electric fields.

Before this can be done, however, a few equations and theories sometimes omitted in elementary courses in physics and physical chemistry must be developed. These include the Boltzmann distribution, the Poisson equation, and the Debye-Hückel theory.

Up until now, no special mention was made of the system of units used to describe physical situations. Actually, it makes little difference in the physics discussed thus far whether one accepts the gram, or the kilogram, as the unit of mass and the centimeter, or the meter, as the unit of length, as is done in the c.g.s. system or the m.k.s. system, for the equations are the same, and it is relatively easy to convert from one system to the other.

In the field of electricity and magnetism, the two systems differ more fundamentally, to the extent that the equations no longer look the same. The physics is still the same and it is still possible to convert from one system to the other, but not quite as easily. It does matter which system one adopts. Half a century ago, the tendency among physicists was to use the c.g.s. system in elementary instruction. Within the past few decades, there has been a shift to the m.k.s. or international system. Physical chemists have followed suit, but more slowly. It is probable, however, that most biologists who studied their physics within the past two or three decades are familiar with the equations expressed in the manner of the m.k.s. system. Furthermore, this system is definitely simpler for computations. These facts, plus the perception that the c.g.s. system is "old-fashioned," argue in favor of the adoption of the m.k.s. system. However, the theoretical work underlying conductance and electrokinetics, the unifying categories in this chapter and the next, is rather old. In most of the reference books that will be cited, the equations are expressed in the c.g.s. form. For this reason, the author has adopted that system, accepting the risk of being thought "old-fashioned." However, to help the reader familiar with electrostatic equations expressed in the m.k.s. system bridge the gap, the author begins with a few paragraphs of translation.

In Table 4.1 a bar above a symbol designates a vector, and symbols without bars are either scalar magnitudes of vectors or scalar quantities. The symbol **D** is the dielectric constant. It is frequently omitted in physics texts, with the understanding that the equation applies to charges in a vacuum where **D** is unity. *E* is electric field strength or electric intensity. In the c.g.s. system it is the force in dynes exerted on a positive charge of 1 statcoulomb. Q is area and q is charge. F is force and r is distance from a point charge or distance between point charges, expressed in the c.g.s. system as dynes and centimeters, respectively. **V** is potential, expressed as statvolts in the c.g.s. system, or as volts in the m.k.s. system. In the c.g.s. system, the change in potential in statvolts is the work that would have to be done on a positive charge of 1 statcoulomb to move it between two potential levels. It is conventional to present Gauss' law as the fundamental empirical law and derive equations 4.2 through 4.5 from equation 4.1. Since such derivations are found in textbooks of general physics, they will not be reproduced here. However, the experimental basis for Gauss' law, equation 4.1, is Coulomb's law, equation 4.3, Gauss' law is deduced from it. The choice of Gauss' law as the fundamental law is in part at least aesthetic. The numerical value of the permittivity constant, ϵ_o, is $(3.336 \times 10^{-3}$ statvolts/volt) (100 cm/m)/[$4\pi(2.998 \times 10^9$ statcoulombs/coulombs)].

The interconversion between the m.k.s. and the c.g.s. systems can be illustrated by a simple hypothetical example. Calculate the potential 1 meter away from the center of a small conducting sphere with a positive charge of 10^{-6} coulombs in a medium (vacuum) in which **D** is 1. By the m.k.s. version of equation 4.4, $\mathbf{V} = 10^{-6}/4\pi(8.85 \times 10^{-12}) = 9 \times 10^3$ volts. By the c.g.s. version, $\mathbf{V} = 10^{-6}(2.998 \times 10^9$ statcoulombs/

TABLE 4.1. Comparison of m.k.s. and c.g.s. Systems

	m.k.s. System	c.g.s. System	Equation
Gauss's law (flux)	$\epsilon_o \oint \bar{E} \cdot d\bar{Q} = q/\mathbf{D}$	$\oint \bar{E} \cdot d\bar{Q} = 4\pi q/\mathbf{D}$	4.1
Electric field strength (electric intensity) surrounding a point charge, q	$E = q/4\pi\epsilon_o\mathbf{D}\, r^2$	$E - q/\mathbf{D}\, r^2$	4.2
Coulomb's law	$F = qq_o/4\pi\epsilon_o\mathbf{D}\, r^2$	$F = qq_o/\mathbf{D}\, r^2$	4.3
Potential in the neighborhood of a point charge, q	$\mathbf{V} = q/4\pi\epsilon_o\mathbf{D}\, r$	$\mathbf{V} = q/\mathbf{D}\, r$	4.4
	$d\,\mathbf{V}/dr = -q/4\pi\epsilon_o\mathbf{D}r^2 = -E$	$d\,\mathbf{V}/dr = -q/\mathbf{D}\, r^2 = -E$	4.5
Definitions	$q \equiv$ coulombs $= 2.998 \times 10^9$ statcoulombs $\mathbf{V} \equiv$ volts $= 3.336 \times 10^{-3}$ statvolts $r \equiv$ meters $= 100$ cm $\epsilon_0 \equiv$ permittivity constant $= 8.85 \times 10^{-12}$ farad/meter	$q \equiv$ statcoulombs $\mathbf{V} \equiv$ statvolts $r \equiv$ cm	
Protonic charge	$\epsilon = 1.602 \times 10^{-19}$ coulombs	$\epsilon = 4.774 \times 10^{-10}$ statcoulombs	

coulomb)/100 cm/m = 29.98 statvolts = 29.98/3.336 × 10^{-3} statvolt/volt = 9 × 10^3 volts.

BOLTZMANN DISTRIBUTION LAW

The Boltzmann distribution law will be used subsequently in the form of equation 4.6:

$$\nu/\nu_0 = e^{-\mathrm{P.E.}/\mathbf{k}T} \tag{4.6}$$

In equation 4.6, ν is the number of molecules per unit volume having a potential energy, P.E., ν_o is the number of molecules per unit volume having 0 potential energy, **k** is the Boltzmann constant or the gas constant per molecule, and T is the absolute temperature. A statistical mechanical derivation of this equation is given by Moore [1]. A derivation that applies specifically to the distribution of gas molecules in a gravitational field can be found in various older books, for example, in MacInnes [2]. Consider a column of area Q containing an ideal gas of mass, *m*, per molecule in a gravitational field, g. Let h represent height. The ideal gas law can be written $P = (n/V)RT = \nu \mathbf{k}T$ and $dP = \mathbf{k}Td\nu$. In these equations, P is pressure, n is number of moles, V is volume, ν is number of molecules per unit volume. As is illustrated in Figure 4.1, an alternative expression for dP contributed by the molecules in the element of volume Qdh can be written $dP = -\nu Qdh\,mg/Q = -\nu mg\,dh$. When these two expressions for dP are equated and variables are collected, $d\nu/\nu = -mg\,dh/\mathbf{k}T$. Upon integration, $\ln(\nu/\nu_0) = -mgh/\mathbf{k}T = -\mathrm{P.E.}/\mathbf{k}T$. When antilogs are taken, equation 4.6 is obtained.

POISSON'S EQUATION

Gauss' law, equation 4.1, represents the scalar product of the electric field strength directed outward and the area surrounding a charge, q. Now imagine an elemental cube such as that in Figure 3.1 surrounding the charge, q. $\bar{E}$ in equation 4.1 is a vector and therefore has components in the x-, y- and z-directions, $\bar{E}_x$, $\bar{E}_y$, and $\bar{E}_z$, respectively. If $\bar{E}_x$ is the electric intensity in the direction left to right at the left face of the elemental cube, then $-\bar{E}_x$ would be the electric intensity directed

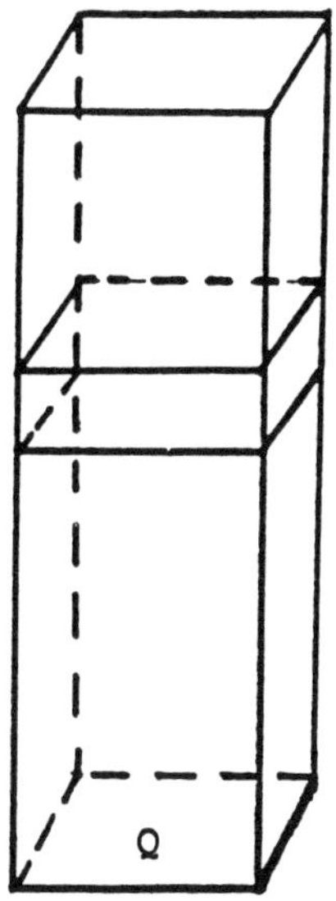

Fig. 4.1. Diagram illustrating a column of gas at equilibrium in a gravitational field. The number of molecules per unit volume and the pressure both decrease from bottom to top.

outward at that point. $\bar{E} \cdot d\bar{Q}$ at that point would then be represented by $E_x \cos 180° \, dydz = -E_x \, dydz$. On the right-hand face, by similar reasoning, the scalar product, $\bar{E} \cdot d\bar{Q}$, is $(E_x + (\partial E_x/\partial x)dx)dydz$. The sum of the two is $(\partial E_x/\partial x)dxdydz$. Similar expressions are obtained in the same manner for the y- and z-directions, and the sum for all six surfaces of the cube is $\oint \bar{E} \cdot d\bar{Q} = (\partial E_x/\partial x + \partial E_y/\partial y + \partial E_z/\partial z)dxdydz$. The right member of equation 4.1 in the c.g.s. form can be written $4\pi \boldsymbol{\rho} \, dxdydz/\mathbf{D}$. In this equation, $\boldsymbol{\rho}$ is the charge density or charge per unit volume. When these expressions for the left and the right members of equation 4.1 are equated, the elemental volume, dxdydz, occurs on both sides and therefore cancels. Thus $(\partial E_x/\partial x + \partial E_y/\partial y + \partial E_z/\partial z) = 4\pi\boldsymbol{\rho}/\mathbf{D}$. By analogy with equation 4.5, $-E_x = \partial \mathbf{V}/\partial x$, etc. Thus,

$$\partial^2 \mathbf{V}/\partial x^2 + \partial^2 \mathbf{V}/\partial y^2 + \partial^2 \mathbf{V}/\partial z^2 = -4\pi \boldsymbol{\rho}/\mathbf{D} \equiv \nabla^2 \mathbf{V} \qquad (4.7)$$

This is Poisson's equation expressed in terms of Cartesian coordinates.

By comparison of equation 3.16 with equation 3.5, it is clear that in radially symmetrical polar coordinates

$$\nabla^2 \mathbf{V} = \partial^2 \mathbf{V}/\partial r^2 + (2/r)\partial \mathbf{V}/\partial r = (1/r)\partial^2(r\,\mathbf{V}/\partial r^2) \qquad (4.8)$$

That the right-hand member of equation 4.8 is equal to the central member can be proved by differentiation. The definition of $\nabla^2\mathbf{V}$ given by the right member of equation 4.8 is the form most useful for development of the Debye-Hückel theory.

DEBYE-HÜCKEL THEORY

When a charged surface is in contact with an electrolyte solution, there must be an excess of oppositely charged ions in the liquid in the neighborhood of the charged surface. Helmholtz [3] thought of this countercharge in the liquid as a layer at a fixed distance from the solid surface; the two charged layers, one on the solid and the opposite in the liquid, became known as the Helmholtz double layer. Gouy [4] and Chapman [5] revised this concept by assuming that the opposite charges in the liquid were distributed in an ionic atmosphere. This idea was developed by Debye and Huckel [6] for the case of a charged spherical particle.

Consider a spherical particle or ion of radius r_1 and charge q, surrounded by an atmosphere of opposite charges and other charges in an electrolyte solution. The atmosphere is illustrated in Figure 4.2. At a great distance from the surface, positive and negative charge densities will be equal, but, close to the surface, charges opposite to that of the surface will outnumber charges of the same sign derived from the electrolyte. Furthermore, the opposite charges will tend to be drawn toward the surface and the charges of the same sign repelled. Close to the surface this attraction and repulsion will be greater than farther away. This is what is meant by the ion atmosphere. Bring an ion of charge $z_i\epsilon$ from where the potential is 0 to a point near the sphere where the potential is **V**. The work involved, that is, the increase in potential energy, is $z_i\epsilon\mathbf{V}$. z_i is the valence of the *i*th ion and ϵ is the protonic charge. The convention is adopted that z_i is positive for cations and negative for anions; some authors define z_i as the absolute magnitude of *z,* always positive. If ν_i is the number of such ions per unit volume where

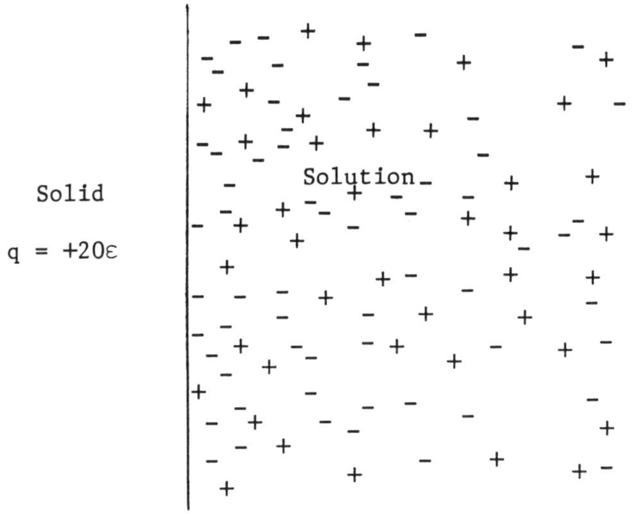

Fig. 4.2. Diagram illustrating the "ionic atmosphere" in a solution in contact with a solid surface with a total positive charge of 20 proton units.

the potential is 0 and ν is the number per unit volume at potential $\mathbf{V}$, then by the Boltzmann distribution law, equation 4.6, $\nu = \nu_i e^{-z_i\epsilon\mathbf{V}/\mathbf{k}T}$. Similar equations apply to all ions present. The charge density, $\boldsymbol{\rho}$, is the algebraic sum for all the ions.

$$\boldsymbol{\rho} = \Sigma\, z_i \epsilon \nu_i e^{-z_i\epsilon\mathbf{V}/\mathbf{k}T} \tag{4.9}$$

If the positive and negative ions are of the same valence, $\pm z$, equation 4.9 can be transformed into equation 4.10, in which ν_o is the number of ions of either type where $\mathbf{V} = 0$.

$$\boldsymbol{\rho} = (\nu_+ - \nu_-)\epsilon = z\epsilon\nu_o(e^{-z\epsilon\mathbf{V}/\mathbf{k}T} - e^{+z\epsilon\mathbf{V}/\mathbf{k}T} = -2\, z\nu_o\, \epsilon \sinh z\epsilon\mathbf{V}/\mathbf{k}T \tag{4.10}$$

When equation 4.10 is substituted into equation 4.7

$$\nabla^2 \mathbf{V} = (8\pi\nu_o\epsilon z/\mathbf{D}) \sinh z\epsilon\mathbf{V}/\mathbf{k}T \tag{4.11}$$

Equation 4.11 actually applies to a surface of any shape. However, problems arise in attempting to integrate it for spherical charged particles. For this reason, Debye and Hückel solved the problem by expanding equation 4.9.

$$\rho = \Sigma\, z_i \epsilon \nu_i (1 - z_i \epsilon \mathbf{V}/\mathbf{k}T + \text{etc.})$$

When $z_i \epsilon \nu_i/\mathbf{k}T$ is much less than 1,

$$\rho = \Sigma\, z_i \epsilon \nu_i - \Sigma\, z_i^2\, \epsilon^2 \nu_i\, \mathbf{V}/\mathbf{k}T$$

This is known as the Debye-Hückel approximation. Because of electrical neutrality where **V** is 0, the first term on the right is 0.

$$\rho = -\Sigma\, \nu_i (z_i \epsilon)^2\, \mathbf{V}/\mathbf{k}T \qquad (4.12)$$

When this is substituted into equation 4.7, Poisson's equation, and the definition of $\nabla^2\mathbf{V}$ appropriate for polar coordination, equation 4.8, is used, equation 4.13 results.

$$(1/r)(\partial^2(r\mathbf{V})/\partial r^2) = 4\pi\epsilon^2\, \mathbf{V}\, (\Sigma\, \nu_i z_i^2)/\mathbf{DkT} \equiv \boldsymbol{\kappa}^2\mathbf{V} \qquad (4.13)$$

$\boldsymbol{\kappa}$ is the famous Debye-Hückel constant.

$$\boldsymbol{\kappa} = \sqrt{4\pi\, \mathbf{A}\, \epsilon^2/1{,}000\, \mathbf{DkT}}\, \sqrt{\Sigma\, c_i z_i^2} \qquad (4.14)$$

In this equation c_i is concentration expressed in moles per liter. $\boldsymbol{\kappa}$ has the dimension of reciprocal length. Ionic strength, μ, is defined as $(1/2)\, \Sigma c_i z_i^2$. In water at 25°C,

$$\kappa = 0.327 \times 10^8 \sqrt{\mu} \qquad (4.15)$$

Because $\sqrt{\mathbf{DT}}$ for water varies only slightly with temperature, equation 4.15 is accurate within ±1% for the temperature range, 10–40°C. Equation 4.13 can now be rewritten as equation 4.16.

$$\partial^2(r\,\mathbf{V})/\partial r^2 = \boldsymbol{\kappa}^2(r\mathbf{V}) \qquad (4.16)$$

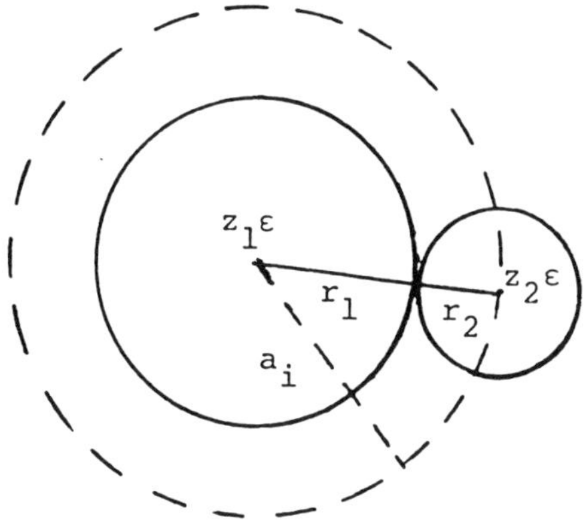

Fig. 4.3. Diagram illustrating the closest distance of approach, $\mathbf{a}_i$, of two charged spheres.

Its solution[1] is $\mathbf{rV} = Ae^{-\kappa r} + Be^{\kappa r}$, where A and B are arbitrary constants.

$$\mathbf{V} = (A/r)e^{-\kappa r} + (B/r)e^{\kappa r}.$$

Since $\mathbf{V}$ is 0 when r is ∞, B must equal 0 because $1/r\ e^{\kappa r}$ increases without limit as r approaches ∞.

$$\mathbf{V} = (A/r)e^{-\kappa r}$$

To evaluate A, consider two spheres of radii r_1 and r_2 as illustrated in Figure 4.3. The distance of closest approach is $\mathbf{a}_i$ equal to $r_1 + r_2$. Let r represent the distance from the center of one of the spheres with a charge, $z_1\epsilon = q$. When $r \geq \mathbf{a}_i$, $\mathbf{V} = (A/r)e^{-\kappa r}$ and $d\mathbf{V}/dr = -(A/r^2)e^{-\kappa r}(1 + \kappa r)$. When $r = \mathbf{a}_i$, $d\mathbf{V}/dr = -A/\mathbf{a}_i^2\ e^{-\kappa \mathbf{a}_i}(1 + \kappa \mathbf{a}_i)$. Two basic facts of physics must now be encountered. One is that a spherically symmetrical distribution of charges outside a sphere makes no contribution to the electric field strength, E, inside the sphere. Since by equation 4.5, $-E = d\mathbf{V}/dr$, this means that the spherically symmetrical distribution of counter charges at $r \geq \mathbf{a}_i$ does not contribute to $d\mathbf{V}/dr$ when $r \leq \mathbf{a}_i$. The second basic fact is that outside a symmetrically charged sphere, E has the same value as it would have if the charge were concentrated at the center. This means that the contribution of the charge, q, on the central sphere to E at $r = \mathbf{a}_i$ is given by equation 4.2. Thus, when $r = \mathbf{a}_i$, $dV/dr = -E = -q/\mathbf{D}\mathbf{a}_i^2$. When $r = a_i$, these two values of $d\mathbf{V}/dr$ are equal.

$$-q/\mathbf{D}\mathbf{a}_i^2 = -(A/\mathbf{a}_i^2)e^{-\kappa \mathbf{a}_i}(1 + \kappa \mathbf{a}_i)$$
$$A = (q/\mathbf{D})e^{\kappa \mathbf{a}_i}/(1 + \kappa \mathbf{a}_i).$$

Therefore, when $r > \mathbf{a}_i$,

$$\mathbf{V} = \left[q/[\mathbf{D}r\,(1 + \kappa \mathbf{a}_i)]\right] e^{-\kappa(r - a_i)} \qquad (4.17)$$

When $r = \mathbf{a}_i$

$$\mathbf{V}_{\mathbf{a}_i} = [(q/\mathbf{D}\mathbf{a}_i)(1/(1 + \kappa \mathbf{a}_i))] = (q/\mathbf{D}\mathbf{a}_i)[1 - \kappa \mathbf{a}_i/(1 + \kappa \mathbf{a}_i)] =$$
$$q/\mathbf{D}\mathbf{a}_i - \kappa q/[\mathbf{D}\,(1 + \kappa \mathbf{a}_i)] \qquad (4.18)$$

Because of the Debye-Hückel approximation, equation 4.12 and equations 4.17 and 4.18 derived from it are only approximations. Actually, equation 4.12 could have been derived from equation 4.10, which is exact, for the case of symmetrical electrolytes by expanding sinh into series and dropping all terms except the first.

$$\sinh x = x + x^3/3! + x^5/5! \text{ etc.}$$

Obviously, for small values of x the first term suffices. When x is 0.5, the error in dropping higher terms is −4%, and when x is 1.0, the error is −15%. Thus the upper limit for which $z\epsilon\,\mathbf{V}/\mathbf{k}T$ is a reasonably accurate approximation of sinh $z\epsilon\,\mathbf{V}/\mathbf{k}T$ is when it does not exceed 1. This is the upper limit of the usefulness of equation 4.12 and of equations 4.17 and 4.18 derived from it. If the highest allowable value of $z\epsilon\,\mathbf{V}/\mathbf{k}T$ is 1, the maximum allowable value of $\mathbf{V}$ is given by $\mathbf{V} = \mathbf{k}T/z\epsilon$. In the c.g.s. system, the Boltzmann constant, $\mathbf{k}$, is 1.38×10^{-16}, and the protonic charge, ϵ, is 4.77×10^{-10} statcoulombs. At 25°C or 298 K, the maximum allowable value of $z\,\mathbf{V}$ is $0.86 \times 10^{-4}/z$ statvolts. Since 1 statvolt = $1/3.336 \times 10^{-3}$ volts or 0.300×10^3 volts, this corresponds to about 0.025 volts or 25 millivolts.

When $4\pi\,\mathbf{a}_i^2\sigma$, where σ is the number of charges per unit area, is substituted for q in the first term on the right in equation 4.18, and when

κa_i is vastly greater than 1, $V_{a_i} = 4\pi\sigma/D\kappa$ and

$$z\epsilon V_{a_i}/kT = 4\pi z\epsilon\sigma/D\kappa kT \quad (4.19)$$

Equation 4.19 should be accurate for values of $z\epsilon V_{a_i}/kT$ of less than unity when κa_i is very much larger than 1. However, when $z\epsilon V_{a_i}/kT$ is greater than 1, that is, when zV_{a_i} is greater than 25 millivolts, this equation is no longer valid. Various attempts have been made over the years to derive equations satisfactory for higher values of V_{a_i}. The most satisfactory solutions are made by computer calculations. Loeb, Wiersema and Overbeek summarized and extended earlier work. The results are presented in Table 4.2, reproduced from Overbeek and Lijklema [7]. It can be seen in the upper left-hand corner that when $z\epsilon V_{a_i}/kT$ is 1 and $\kappa a_i = 20$, the result differs from that of equation 4.19 by only 9%. The table provides solutions of equation 4.10 for potentials much higher than 25 millivolts. The enormous deviations from equation 4.19 for the higher values of V_{a_i} are illustrated by comparing the second and third columns.

ACTIVITY COEFFICIENT OF ELECTROLYTES

The important role played by ions in osmotic pressure, and thus in the movement of fluids by osmosis, was alluded to in Chapter 1. The osmotic properties of real solutions depend on the activities of the solutes. As was pointed out in Chapter 1, activity is a measure of escaping tendency. It can be determined directly by measuring the partial pressure of the vapor in equilibrium with a solute in solution. By Henry's law, the partial pressure of the vapor is directly proportional to the activity, a, for nonionizing solutes. In one mode of defining activity, $a = \gamma m$, where γ is the activity coefficient and m is the molality of the solute. As m approaches 0, γ approaches 1, and a equals m. For electrolytes, however, the situation is different. Consider for example the gas, HCl, in equilibrium with an aqueous solution. HCl is in the molecular form in the gas phase but it is ionized in solution. One can describe the equilibrium by the equation, $(a_H)(a_{Cl^-})/P_{HCl} = K$, where P_{HCl} is the partial pressure of the HCl gas and K is the equilibrium constant. Thus the partial pressure of the vapor is proportional to the product of the activities of the two ions. Conceptually, one can think of the activity of each ion as being equal to its activity coefficient

TABLE 4.2. Values of $4\pi z\epsilon\sigma/D\kappa kT$ for Various Values of κa_i and $z\epsilon V_{a_i}/kT$

approx. zV_{a_i} millivolts	$\frac{z\epsilon V_{a_i}}{kT}$	κa_i							
		20	10	5	2	1	0.5	0.2	0.1
25	1	1.0913	1.1403	1.2386	1.5344	2.0297	3.0238	6.0155	11.010
50	2	2.4430	2.5359	2.7230	3.2910	4.2522	6.2019	12.132	22.087
75	3	4.3861	4.5144	4.7740	5.5698	6.9372	9.7562	18.495	33.322
100	4	7.4067	7.5614	7.8744	8.8428	10.530	14.080	25.376	44.896
125	5	12.271	12.444	12.791	13.878	15.785	19.878	33.310	57.170
150	6	20.218	20.403	20.772	21.929	23.965	28.377	43.327	70.864
175	7	33.273	33.465	33.848	35.041	37.138	41.678	57.395	87.544
200	8	54.773	54.967	55.359	56.569	58.677	63.214	79.008	110.34
225	9	90.202	90.397	90.794	92.013	94.106	98.580	113.95	144.95
250	10	148.60	148.80	149.20	150.42	152.50	156.87	171.58	201.08
275	11	244.89	245.09	245.48	246.70	248.77	253.03	267.09	294.58
300	12	403.63	403.83	404.23	405.44	407.49	411.68	425.18	450.71
325	13	665.34	665.53	665.94	667.14	669.17	673.33	686.38	710.29
350	14	1096.8	1097.0	1097.4	1098.6	1100.6	1104.8	1117.5	1140.2
375	15	1808.2	1808.4	1080.8	1810.0	1812.0	1816.1	1828.6	1850.5
400	16	2981.2	2981.4	2981.8	2983.0	2985.0	2989.0	3001.3	3022.5

multiplied by the molality, m, of HCl in solution. Thus the partial pressure is proportional to $\gamma_+ \gamma_- m^2$. Operationally, one can measure only the mean activity coefficient, $\gamma_\pm$ from vapor pressure measurements or, indeed, from any other kind of measurement. In a vapor pressure experiment, the partial vapor pressure is proportional to $(\gamma_\pm m)^2$. It is evident that measurable $\gamma_\pm^2$ equals conceptual $\gamma_+ \gamma_-$ or $\gamma_\pm = \sqrt{\gamma_+ \gamma_-}$. A more general expression for ions of any valence type is

$$\gamma_\pm = (\gamma_+^{\nu_+} \gamma_-^{\nu_-})^{1/(\nu_+ + \nu_-)} \tag{4.20}$$

In this equation, ν_+ and ν_-, respectively, are the numbers of equivalents of positive and negative ion formed by the dissociation of one mole of the electrolyte. Even very dilute solutions of electrolytes are nonideal. It is possible to derive a theoretical equation for the activity coefficient of an ion from Debye-Hückel theory.

The chemical potential of the *i*th ion, μ_i, equals $\mu_{io} + RT\ln a_i =$

$$\mu_{io} + RT\ln N_i + RT\ln \gamma_i = \mu_i\,(\text{ideal}) + RT\ln \gamma_i.$$

Now, make the assumption that the departure from ideality of an ion is entirely the result of electrostatic interaction; in other words, ignore all contributions to nonideality except electrostatic interaction. For the *i*th ion, one can now write $G_i = G_i\,(\text{ideal}) + G_{ei}$. G_{ei} is the electrostatic interaction contribution to G_i.

$\partial G_i/\partial n_i = \partial G_i\,(\text{ideal})/\partial n_i + \partial G_{ei}/\partial n_i$, or $\mu_i = \mu_i\,(\text{ideal}) + \partial G_{ei}/\partial n_i$.
From these two equations for μ_i, it follows that $\partial G_{ei}/\partial n_i = RT\ln \gamma_i$.
When ν_i is the number of ions of the *i*th sort, $\partial G_{ei}/\partial \nu_i = (RT/\mathbf{A})\ln\gamma_i = \mathbf{k}T\ln\gamma_i$

(4.21)

Inspection of the extreme right of equation 4.18 shows that $\mathbf{V}_{\mathbf{a}_i}$ is composed of two terms. The first is $q/\mathbf{D}\mathbf{a}_i$, which is the contribution to the potential from the charge on the sphere itself and is, thus, independent of the concentration of surrounding ions. The second contribution is $-\boldsymbol{\kappa} q/[\mathbf{D}(1 + \boldsymbol{\kappa}\mathbf{a}_i)]$. This is the contribution from the interaction of the ions in the solution; it is concentration-dependent because $\boldsymbol{\kappa}$ is a function of ionic strength. By definition, nonideality is concentration-dependent. Therefore, the departure from ideality is

TABLE 4.3. Debye-Hückel Constants for Aqueous Solutions

°C	A	$B \times 10^{-8}$
0	0.4863	0.3243
18	0.4992	0.3272
25	0.5056	0.3286
38	0.5186	0.3314

dependent on the second contribution to $\mathbf{V}_{\mathbf{a}_i}$, $\mathbf{V}' = -\kappa q/[\mathbf{D}(1 + \kappa \mathbf{a}_i)]$. The contribution to the Gibbs free energy of a charged particle is equal to the work of charging that particle from 0 to its final charge and final potential. As proposed by Güntelberg [8] and by Müller [9], the contribution resulting from the influence of surrounding ions on the free energy of a single ion, $\partial G_e/\partial \nu_i = \int_0^q \mathbf{V}'\, dq$. Thus,

$$\mathbf{k}T\ln\gamma_i = \int_0^q -[\kappa q/\mathbf{D}(1 + \kappa \mathbf{a}_i)]\, dq = -\kappa q^2/2\mathbf{D}(1 + \kappa \mathbf{a}_i)$$

Since for an ion, q is $z_i\, \epsilon$,

$$-\ln \gamma_i = \left[z_i^2 \epsilon^2/2\, \mathbf{D}\mathbf{k}T\right]\left[\kappa/(1 + \kappa \mathbf{a}_i)\right] \qquad (4.22)$$

When the definition of κ is inserted, and the logarithm is converted to the base 10,

$$-\log \gamma_i = z_i^2 A\sqrt{\mu}/(1 + B\, \mathbf{a}_i \sqrt{\mu}) \qquad (4.23)$$

The values for A and B for aqueous solutions at several different temperatures are given in Table 4.3.

Equations 4.22 and 4.23 are theoretical expressions for γ_i, the activity coefficient of a single ion. These equations cannot be put to experimental test because there is no way of measuring the activity coefficient of an ion. However, as was mentioned previously, the mean activity coefficient, $\gamma_\pm$, can be measured by any thermodynamic method that depends upon the activity. Among these are the colligative properties—vapor pressure, freezing point depression, boiling point elevation, and osmotic pressure—solubility measurements, and electrochemical measurements. When a is determined for various measured values of m, $\gamma_{\pm}$ can be determined as a function of m.

If equation 4.23 is substituted into equation 4.20 expressed in logarithmic form, equation 4.24 is obtained.

$$-\log \gamma_{\pm} = |z_+ z_-| A \sqrt{\mu}/(1 + B \mathbf{a}_i \sqrt{\mu}) \qquad (4.24)$$

The algebraic simplification[2] is possible because electrical neutrality requires that $|\nu_+ z_+| = |\nu_- z_-|$.

The Debye-Hückel theory and equation 4.24 deduced from it have been subjected to extensive theoretical criticism. The various mathematical approximations made in its derivation are not the only objections to have been raised. The $\mathbf{a}_i$ in equation 4.24 is an average distance of closest approach of ions in an electrolyte solution. It is in a sense arbitrary because it is customary to choose a value that gives the best fit of theoretical and experimental values of log $\gamma_{\pm}$. This is not a serious objection if the value chosen is at least approximately the same as the distance between ions in the crystal lattice.

Despite all objections, the Debye-Hückel theory as expressed in equation 4.24 is remarkably successful in predicting the activity coefficient of electrolytes in dilute solutions. There is almost perfect agreement between calculated and experimentally determined values of $\gamma_{\pm}$ for KCl up to concentrations of 0.10 m. For many electrolytes, calculated and experimental values differ by only a few per cent up to concentrations two or three times that high. For more concentrated solutions, equation 4.24 invariably fails. Semiempirical corrections for a "salting-out effect" have been proposed. Such a modified equation 4.24 agrees with experiments up to considerably higher concentrations. In biological systems, however, electrolyte concentrations are usually low enough to justify the use of equation 4.24 to calculate $\gamma_{\pm}$.

CONDUCTIVITY OF ELECTROLYTE SOLUTIONS

The starting point for a consideration of the conductivity or conductance of electrolyte solutions is Ohm's law, equation 4.25.

$$\mathbf{V} = \mathbf{RI} \qquad (4.25)$$

In this equation, **V** is the potential expressed in volts, **I** is the current in amperes or coulombs per second, and **R** is the resistance expressed in ohms. Specific resistance is the resistance of a column 1 cm long and 1 cm in cross section. Conductivity or conductance is the reciprocal of resistance expressed in reciprocal ohms or mhos, and specific conductance, **Λ**, is the conductance of a solution between electrodes 1 cm^2 in area 1 cm apart. Equation 4.25 can thus be modified to give equation 4.26.

$$\mathbf{\Lambda} = \mathbf{IL/VA} \tag{4.26}$$

Ideally, specific conductance should be directly proportional to concentration. Actually, this is true only at very low concentrations in real solutions. Equivalent conductance is the conductance in mhos of a solution containing one equivalent of electrolyte between two electrodes one cm apart. Conceptually, the area of the electrodes must be large enough to accommodate the solution containing one equivalent, no matter what the volume. However, equivalent conductance, Λ, can be determined from the specific conductance, **Λ**, by multiplying by the volume containing one equivalent.

$$\Lambda = 1{,}000\ \mathbf{\Lambda}/c \tag{4.27}$$

In this equation, c is the concentration in equivalents per liter, the normality. Since current is carried by the ions and the number of ions is independent of dilution in a fully ionized solution containing one equivalent, ideally equivalent conductance, Λ, should be independent of concentration. However, in practice it decreases with increased concentration. The decrease can be quite significant.

The equivalent conductance of an electrolyte is the sum of the equivalent conductances of the cations and anions, λ_+ and λ_-, respectively. The transference number, *t,* is the ionic conductance of an ion divided by the sum of the ionic conductances of all the ions, as shown in equation 4.28.

$$t_+ = \lambda_+/(\lambda_+ + \lambda_-) \quad ; \quad t_- = \lambda_-/(\lambda_+ + \lambda_-) \tag{4.28}$$

Mobility, u, is defined as the velocity of an ion or charged particle per unit of potential gradient. The transference number of an ion is also equal to the mobility of that ion divided by the sum of the mobilities of all ions. The transference number is in reality the fraction of the total current carried by a particular ion.

Decrease in equivalent conductance, Λ, with increasing concentration requires further discussion. For weak electrolytes the ratio of Λ, the equivalent conductance at a particular concentration, to Λ_0, the value at infinite dilution, is equal to the degree of dissociation of the weak electrolyte. This is the basis of the original Arrhenius method for determining degree of dissociation. Originally, it was thought that strong electrolytes, while largely dissociated, were also partially undissociated in concentrated solutions, and this same ratio was explained in terms of degree of dissociation. For many reasons, the modern view is that strong electrolytes are completely dissociated even at high concentrations. The theoretical explanation for the decrease of equivalent conductance of strong electrolytes as concentrations increase stems in part from the Debye-Hückel theory and in part from a statistical mechanical theory developed by Onsager [10]. The part related to the Debye-Hückel theory is called the electrophoretic effect, and the part derived by statistical mechanics is called the time of relaxation effect.

Electrophoretic effect is in reality a misnomer. As will be explained in the next chapter, the effect here described is an electro-osmotic effect, because it involves the bulk movement of liquid containing ions.

ELECTROPHORETIC EFFECT

The electrophoretic effect is derived in the following manner. If in the Debye-Hückel approximate expression of $\boldsymbol{\rho}$, equation 4.12, the concentration term is converted from ions per unit volume, ν_i, to moles per liter, c_i, and then if the definition of $\boldsymbol{\kappa}$, equation 4.13., is substituted,

$$\boldsymbol{\rho} = -\mathbf{D}\boldsymbol{\kappa}^2\mathbf{V}/4\pi.$$

If now equation 4.17 for the total potential at $r \geq \mathbf{a}_i$, where $z_i\epsilon$ replaces q, is substituted,

$$\boldsymbol{\rho} = -\left[z_i\epsilon\boldsymbol{\kappa}^2\, e^{\boldsymbol{\kappa}\mathbf{a}_i}/4\pi(1 + \boldsymbol{\kappa}\mathbf{a}_i)\right] e^{-\boldsymbol{\kappa}r}/r \equiv -Ae^{-\boldsymbol{\kappa}r}/r \qquad (4.29)$$

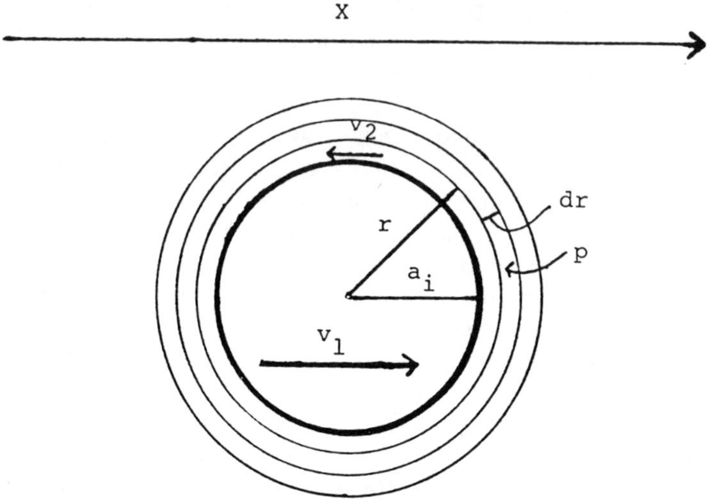

Fig. 4.4. Cross-sectional diagram of a charged sphere with radius of closest approach of $\mathbf{a}_i$ surrounded by concentric shells of liquid of radius r and thickness dr. In each shell the charge density is ρ, which has its maximum value in the shell at $r = \mathbf{a}_i$ and decreases to zero at $r = \infty$. The long vector at the top represents the electric field. The vector v_1 represents the velocity of the sphere with respect to liquid, and v_2 is the velocity of the liquid adjacent to the sphere with respect to stationary liquid at great distance from the sphere.

Imagine as illustrated in Figure 4.4 concentric spherical shells of radius r and thickness dr in the liquid surrounding the charged sphere. In each shell the total charge will be its volume multiplied by $\boldsymbol{\rho}$. Electrostatic force is charge times the imposed potential gradient, X. Since $\boldsymbol{\rho}$ increases from zero at $r = \infty$ to maximum value when $r = \mathbf{a}_i$, the electrostatic force on the successive shells increases from zero to a maximum at $r = \mathbf{a}_i$. The increment in force between two successive shells, dF, is given by $dF = Xdq = 4\pi r^2\, dr\, \boldsymbol{\rho}X$. When this force is exerted on the ions in the spherical shell, the entire shell and all of the liquid inside it will move with a constant velocity. The reason for this has nothing to do with ionic hydration, as is suggested by some writers; it is purely hydrodynamic. If there were any movement of the ions with respect to liquid in or inside the shell, an accelerating force would be exerted on that liquid until the velocity was the same as that of the ions. Each successive spherical shell will move with a velocity dv_2 relative to its neighboring shell. By Stokes' law, equation 2.15, $dF = 6\pi r\eta dv_2$. When these two expressions for dF are equated and solved for dv_2 and when equation 4.29 is substituted for ρ, $dv_2 = (2/3\eta)X\rho r dr = -(2/3\eta)XAe^{-\kappa r}dr$. The outermost shell at $r = \infty$ will have a velocity of zero and the innermost at $r = a_i$ will have an accumulated velocity of v_2 given by

$$v_2 = \int_0^{v2} dv_2 = \int_\infty^{ai} -(2/3\eta)\, X\, Ae^{-\kappa r}\, dr = (2/3\eta\kappa)\, X\, Ae^{-\kappa a_i}$$

When the definition of A from equation 4.29 is introduced,

$$v_2 = z_i\, \epsilon X\kappa/6\pi\eta(1 + \kappa a_i) \simeq z_i\, \epsilon X\kappa/6\pi\eta \text{ when } \kappa a_i << 1 \quad (4.30)$$

If it were not for the ionic atmosphere, the velocity of the central ion would be v_1. Since the motion of the liquid resulting from the atmosphere is in the opposite direction, the central ion bucks the stream, and the net velocity, v, is

$$|v| = |v_1| - |v_2| = |v_1| - |z_i|\, \epsilon X\kappa/6\pi\eta \quad (4.31)$$

The mobility of an ion or charged particle, u, is v/X. Thus, for each ion, $u = u_1 - z_i\, \epsilon\kappa/6\pi\eta$.

$$\Lambda = \mathbf{F}\,(\nu_+\, |z_+|\, u_+| + \nu_-|z_-|\, |u_-|\,).$$

F is the Faraday constant. Thus,

$$\Lambda = \Lambda_0 - \mathbf{F}\,(\nu_+\, z_+^2\, \epsilon\kappa/6\pi\eta + \nu_-\, z_-^2\, \epsilon\kappa/6\pi\eta)$$

When the definition of κ is inserted in equation 4.15, and all constants are collected into B,

$$\Lambda = \Lambda_0 - B\sqrt{\mu} \quad (4.32)$$

For univalent ions in water at 25°C, B has a value of 59.78. MacInnes [2] has made a numerical calculation that shows that 63% of the decrease in equivalent conductance of KCl in water at 25°C between zero and 0.001 normal is accounted for by this effect.

TIME OF RELAXATION EFFECT

The second effect, the time of relaxation effect, is derived by statistical mechanics and will not be presented in detail here. The gist of the idea is that a stationary charged sphere can be assumed to be surrounded by a spherically symmetrical atmosphere of opposite charges. The electrostatic center of the opposite charges will coincide with the electrostatic center of the charged sphere. When the charged sphere moves, the countercharges in the atmosphere must readjust. While this readjustment is rapid, it is not instantaneous. Thus, for a constantly moving charged sphere, the electrostatic center of charge of the atmosphere lags behind that of the charged sphere and imposes on the sphere an electrostatic force in the direction opposite to the motion. The statistical mechanical derivation shows that this retarding force is proportional to Λ_o and inversely proportional to $(\mathbf{D}T)^{3/2}$, but, unlike the electrophoretic effect, independent of viscosity. Thus an additional term must be added to equation 4.32 to take into account the time of relaxation. This can be presented in simple form as equation 4.33.

$$\Lambda = \Lambda_0 - (A\Lambda_0 + B)\sqrt{\mu} \qquad (4.33)$$

For aqueous solutions of univalent ions at 25°C, A has a value of 0.229. Equation 4.33 is frequently called the Onsager equation. There seems to be a consensus that it fits experimental observations within a small experimental error up to a concentration of about 10^{-3} moles per liter. At somewhat higher concentrations, there is usually a slight but definite deviation.

IONIC MOBILITY

When an electric current is carried by a solution of ions, the positive and negative ions move independently. The positive ions migrate toward the negative electrode and the negative ions toward the positive electrode. Electrical neutrality for the solution is maintained by the electrochemical reactions at the two electrodes. As was mentioned previously, the equivalent conductance of the electrolyte is the sum of the separate ionic conductances, λ_+ and λ_-, and the transference number, t_i, is equal to $\lambda_i/(\lambda_+ + \lambda_-)$. Alternatively, t_i is equal to λ_i/Λ. Transference numbers can be measured by three different methods: 1) the Hittorf method, which involves making analyses of the concentra-

TABLE 4.4. Ionic Mobility at Infinite Dilution in Water at 25°C

Ion	u	r(Å)	Ion	u	r
H^+	36.2×10^{-4}		OH^-	20.5×10^{-4}	
Cs^+	8.0×10^{-4}	1.69	Br^-	8.13×10^{-4}	1.95
K^+	7.61×10^{-4}	1.33	I^-	7.97×10^{-4}	2.16
$NH_4{}^+$	7.60×10^{-4}	1.48	Cl^-	7.92×10^{-4}	1.81
Na^+	5.20×10^{-4}	0.95	$NO_3{}^-$	7.40×10^{-4}	
Li^+	4.01×10^{-4}	0.60	CH_3COO^-	4.25×10^{-4}	
Ca^{2+}	6.16×10^{-4}	0.99	$SO_4{}^{2-}$	8.28×10^{-4}	

tion of ions in the neighborhood of the electrodes after electrolysis; 2) the moving boundary method, which resembles somewhat the moving boundary method in electrophoresis; and 3) the electromotive force method. It is not the purpose of this book to discuss physicochemical methods; they are mentioned merely to support the claim that transference numbers can be and have been measured for many ions. Since transference numbers can be measured, ionic conductances, λ_i, can be calculated by the above formula. λ_i is simply $\mathbf{F}u_i$, the Faraday constant times the mobility of the ion, which in turn is equal to the velocity of the ion divided by the potential gradient, X. Ionic mobilities can be measured directly by a method resembling the moving boundary method in electrophoresis. In favorable cases, the mobilities calculated from transference numbers and those measured directly are in reasonable agreement. Table 4.4 shows the mobilities of several ions at infinite dilution in water at 25°C, expressed as cm/sec/volt/cm. The table also shows ionic radii for some of the ions, as determined from crystal structure data.

Since the force exerted on an ion in a potential gradient is $z_i\epsilon X$, for a univalent ion in unit potential gradient, this force can be calculated to be 1.60×10^{-12} dyne. If the ion can be considered to be spherical and if Stokes' law, equation 2.15, applies, the force of resistance, which is equal to the electrical force, is given by $6\pi\eta rv$. In unit potential gradient, X is one and v is equal to u. Thus the mobility, u, should be $\epsilon z_i/6\pi\eta r$. Bull [11] has calculated from this formula that the mobility of a sodium ion should be 10×10^{-4} cm/sec/volt/cm. This figure is approximately double that shown in the Table 4.4.

There are two probable contributing reasons for this remarkable disagreement. One is that Stokes' law might no longer be applicable to molecules that do not differ very much in size from solvent molecules. A second probable contributing factor is electrostriction. The idea behind electrostriction is that water molecules, being dipoles, are oriented in the neighborhood of an ion in such a way that, in the case of a negative ion, the positive pole of the water is oriented toward the ion. Because of the rotational kinetic energy of the water molecules, the orientation is only a statistical average. Since the positive pole of the water will on the average be closer to the negative ion than the negative pole, there will be a net attraction of water toward the ion. When the ion has a positive charge, the average orientation of the water molecule will be with the negative pole toward the ion, but there will still be a net attraction. This net attraction pulls the water molecules closer to the central ion and closer to each other than they would be far from the ion, where all water molecules can be considered to be randomly oriented. A simple experimental observation attests to the reality of electrostriction. When a salt like potassium chloride is dissolved in water, the volume of the solution is less than the sum of the volumes of the pure water and of the added potassium chloride crystals. This is expressed in physicochemical language by saying that the partial specific volume, the derivative of the volume of the solution with respect to the mass of KCl added, is only about seven-tenths of the specific volume (reciprocal of the density) of the solid salt. One can imagine, therefore, that ions are hydrated and that the radius of the hydrated ion is greater than that of the ion in a crystal. There is in fact evidence that when ions migrate, they carry water molecules with them. However, if the idea that electrostriction accounts fully for the mobility of an ion being lower than it should be as calculated from Stokes' law is to be tested, it is necessary to know how much water is firmly bound to an ion. This cannot be determined directly from electrostriction measurements. It is possible to measure the difference between the amounts of waters carried by positive and negative ions during conductivity studies. Values for hydration of individual ions have been calculated from such measurements, but they always depend on an assumption concerning the extent of hydration of one particular ion, the potassium ion.

One must consider then whether it is realistic to use Stokes' law to calculate the frictional coefficient of a small ion. Water molecules, obviously not spheres, have their greatest dimension approximately equal to the diameter of a bromide ion. The derivation of Stokes' law in Chapter 2 involved the simplifying assumption that the moving sphere is large enough compared with the solvent molecules so that the solvent could be treated, by contrast, as a continuous medium. Obviously, small ions or small molecules do not fit this requirement. Yet the use of

Stokes' law does lead to reasonable results for some small molecules. The glycerin molecule, for example, if it were spherical, can be calculated to have a radius of 3.07×10^{-8} centimeters, just a little more than one and a half times the size of a water molecule. The partial specific volume of glycerin is only a few per cent lower than the specific volume; thus the interaction with water molecules is much less than in the case of ions. When the diffusion coefficient of glycerin is calculated from the combination of the Einstein-Sutherland equation, equation 3.4, and Stokes' law, equation 2.16, a value is obtained that is only about 25% higher than the measured coefficient of diffusion of glycerol into water. Since the glycerin molecule isn't a sphere, it should actually diffuse somewhat more slowly than the rate indicated by a calculation based on the assumption that it is a sphere.

The unusually high ionic mobilities of the H^+ and the OH^- ions require further discussion. First of all, free protons do not exist in aqueous solution. They attach themselves to water molecules to give the ion H_3O^+. This ion should be a little larger than a bromide ion, perhaps about the size of an iodide ion, and it should therefore have roughly the same mobility as the bromide or iodide ion. Its much greater mobility can be explained on the assumption that the mode of current flow with hydrogen ions in solution does not involve primarily the actual mobility of H_3O^+ ions but instead the jumping of the proton from one water molecule to the next, a much faster process. Similarly, the exceptionally high value for the mobility of the hydroxyl ion can be explained by the jumping of protons from water molecules to adjacent OH^- ions in the direction opposite the flow of negative charge.

DIFFUSION OF SALTS

The effect of an electric field on the motion of ions has an important influence on the diffusion of salts. In this case, anions and cations must move in the same direction. In the first paragraph of Chapter 3 it was pointed out that the force in dynes per solute molecule, when the only force comes from a gradient of osmotic pressure, is $-(d\pi/dx)/c\mathbf{A}$, when c is moles of solute per unit volume of solvent. When a salt diffuses, the individual ions move separately. To preserve electrical neutrality, they must move at the same rate. Since their tendency is to move at different rates, a potential gradient is established between them. Thus, for the positive and negative ions,

$$F_+ = -(d\pi/dx)/c\mathbf{A} + z_+\epsilon X$$

$$F_- = -(d\pi/dx)/c\mathbf{A} + z_- \epsilon X$$

Since mobility, u, is velocity per unit of X or velocity per dyne, the velocity of each ion, v, is given by

$$v = u_+ [(-d\pi/dx)/c\mathbf{A} + z_+\epsilon X] = u_- [(-d\pi/dx)/c\mathbf{A} + z_- \epsilon X]$$

The middle and right members of this equation can both be solved for ϵX, which is the same for both ions. Thus the two solutions can be equated.

$$(1/\boldsymbol{\nu}_+ z_+)[(v/u_+) + (d\pi/dx)/c\mathbf{A}]\boldsymbol{\nu}_+ = (1/\boldsymbol{\nu}_- z_-)[(v/u_-) + (d\pi/dx)/c\mathbf{A}]\boldsymbol{\nu}_-$$

Since $\boldsymbol{\nu}_+ z_+/\boldsymbol{\nu}_- z_- = -1$, a condition necessary to preserve electrical neutrality,

$$[(v/u_+) + (d\pi/dx)/c\mathbf{A}]\boldsymbol{\nu}_+ = -[(v/u_-) + (d\pi/dx)/c\mathbf{A}]\boldsymbol{\nu}_-$$

and

$$v = -\frac{u_+ u_- (\boldsymbol{\nu}_+ + \boldsymbol{\nu}_-)(d\pi/dx)}{(\boldsymbol{\nu}_+ u_- + \boldsymbol{\nu}_- u_+)\,\mathbf{A}c}$$

By analogy to the derivation of equation 3.1,

$$d\mathbf{S} = c\ v\ Q\ dt = -\frac{u_+ u_- (\boldsymbol{\nu}_+ + \boldsymbol{\nu}_-)}{(\boldsymbol{\nu}_+ u_- + \boldsymbol{\nu}_- u_+)\mathbf{A}}(d\pi/dx)Q\ dt$$

By equation 3.3, $d\mathbf{S} = -DQ(dc/dx)\ dt$. Thus,

$$D(dc/dx) = \frac{u_+u_- (\nu_+ + \nu_-)}{(\nu_+u_- + \nu_-u_+) A} (d\pi/dx)$$

From equation 1.2 and the discussion immediately following it,

$$d\pi \equiv dP = -(RT/\bar{V}_1)d \ln a_1 = (n_2/n_1\bar{V}_1) RT d \ln a_2 = c RT d \ln a_2$$

In this equation, c is defined as moles of solute per unit volume of solvent.

$$d\pi = c RT(d \ln c + d \ln \gamma^{\pm}) = RT dc + RT dc(c d \ln \gamma^{\pm}/dc)$$

$$d\pi/dx = RT(1 + d \ln \gamma^{\pm}/d \ln c) dc/dx$$

$$\therefore \quad D = \frac{(u_+u_-)(\nu_+ + \nu_-)}{\nu_+u_- + \nu_-u_+} \mathbf{k} T \left(1 + \frac{d \ln \gamma^{\pm}}{d \ln c}\right) \qquad (4.34)$$

In equation 4.34, c can be expressed in any convenient manner, since d ln c is independent of the way c is expressed, as long as it is approximately proportional to moles per unit volume. For an ideal solution, d ln $\gamma^{\pm}$/d ln c = 0 and D = D_0. Thus

$$D/D_o = (1 + d \ln \gamma^{\pm}/d \ln c) \qquad (4.35)$$

Equation 4.24 can be written for a single dilute uni-univalent electrolyte when $B\mathbf{a}_i \sqrt{\mu} << 1$,

$$-\ln \gamma^{\pm} \simeq 2.303 A \sqrt{c} \qquad d \ln \gamma^{\pm}/d \ln c = -2.303 Ac\, d \sqrt{c}/dc$$

$d \ln \gamma^{\pm}/d \ln c = -2.303 A \sqrt{c}/2 = -0.58 \sqrt{c}$ for an aqueous solution at 25°C.

This implies that in a solution that is 0.01 molar, $D/D_0 = 1 - 0.058 = 0.94$. This calculated result is in approximate agreement with experiment for some uni-univalent electrolytes. The important point is that diffusion coefficients of electrolytes are highly sensitive to concentration. The situation is complex for the diffusion of one salt in the presence of other electrolytes because there is strong coupling between the various ions present.

REFERENCES

1. Moore, W. J. Physical Chemistry, Fourth edition. NJ: Prentice-Hall, Englewood Cliffs, 1972.
2. MacInnes, D. A. The Principles of Electrochemistry. New York: Reinhold, 1939.
3. Helmholtz, H. Wied. Ann. *7*:337, 1879.
4. Gouy, L. J. Phys. *9*:457, 1910.
5. Chapman, D. L. Phil. Mag. (6) *25*:475, 1913.
6. Debye, P. and Hückel, E. Physik. Z. *24*:185, 1923.
7. Overbeek, J. Th. G. and Lijklema, J. In Electrophoresis, M. Bier, (ed.). New York: Academic Press, 1959.
8. Güntelberg, E. Z. Phys. Chem. *123*:199, 1926.
9. Müller, H. Physik, Z. *28*:324, 1927.
10. Onsager, L. Physik. Z. *27*:388, 1926, and *28*:277, 1928.
11. Bull, H. B. An Introduction to Physical Biochemistry, Second Edition. Philadelphia: F. A. Davis Co., 1971.

NOTES

1. This can be proved by differentiation

$$\partial(\mathrm{rV})/\partial \mathrm{r} = Ae^{-\kappa r}(-\kappa) + Be^{\kappa r}(\kappa)$$

$$\partial^2(\mathrm{rV})/\partial \mathrm{r}^2 = Ae^{-\kappa r}(-\kappa)(-\kappa) + Be^{\kappa r}(\kappa)(\kappa) = \kappa^2(Ae^{-\kappa r} + Be^{\kappa r}) = \kappa^2(\mathrm{rV})$$

2. Since

$$|\nu_+ z_+| = |\nu_- z_-|,$$

$$(\nu_+ z_+^2 + \nu_- z_-^2)/(\nu_+ + \nu_-) = (|z_+ z_-|\,\nu_- + |z_- z_+|\,\nu_+)/(\nu_+ + \nu_-) = |z_+ z_-|$$

5

Electrokinetic Phenomena

Half a century ago there was great interest in electrophoresis, the movement of charged macromolecules through a stationary liquid in an electrical potential gradient, as a method for separating proteins and other charged macromolecules. Much of what is known about the proteins of blood plasma was derived from such experiments. In more recent decades, interest has shifted from the movement of charged particles in liquid media to the movement through gels or through liquid-porous solid systems. This has proven to be highly useful for all kinds of separations of charged particles. This, too, is called electrophoresis. It is the author's conviction that this is an inappropriate and confusing choice of name because the actual velocity of migration of charged particles through a stationary porous solid filled with liquid is affected by many factors, only one of which is electrophoresis, the movement of charged particles in a stationary liquid. The focus in this chapter will be on true electrophoresis and on related phenomena, all involving the relative motion of charged solids to adjacent liquid. There are four such phenomena:

1. Electrophoresis, defined as the movement of charged particles in a potential gradient in a stationary liquid.
2. Electroosmosis, defined as the movement of liquid in a potential gradient through pores having a surface charge.
3. Streaming potential, defined as the potential generated by forcing a liquid through a capillary or pores having a surface charge.
4. Sedimentation potential or Dorn effect, defined as the potential generated by sedimentary charged particles.

Freundlich [1] applied the name electrokinetics to this category. There are many possible applications of electrokinetics in biological systems. The movement of blood through vessels with charged walls should generate streaming potentials. Where potential gradients exist, charged macromolecules should move by electrophoresis. Where such gradients exist across charged pores, electroosmosis should occur.

Electrokinetics is an old science. The early history of observations and theory is reviewed by Abramson [2]. Soon after the discovery of the voltaic pile, Reuss, a physician in the Russian court, demonstrated electroosmosis. He passed an electric current through water in a U-shaped tube, the bottom of which was filled with sand. The water level rose in one arm and fell in the other. Additional important reference books include Abramson, Moyer and Gorin [3] and two volumes edited by Bier [4,5].

THEORY OF ELECTROOSMOSIS

Helmholtz [6] provided the first quantitative theory for electroosmosis. He assumed a uniform distribution of charges on the solid surface and a layer of equal magnitude of opposite charges in the liquid, separated from the solid by a distance, d. This is known as the Helmholtz double layer. If the stationary solid is a nonconducting tube of circular cross section with $r >> d$, when a potential gradient, X, is applied in the direction of the length of the tube, the charged layer in the liquid will move toward the electrode of opposite sign. Since all of the liquid in the tube, except the trivial amount within the double layer, will be surrounded by the moving layer of charges, it will move with the same velocity as the charges. However, this is not a realistic picture of the nature of the double layer. As was discussed in Chapter 4, Gouy [7] pointed out that a much more realistic picture is an atmosphere of opposite charges in the double layer region. The derivation of the equation for eletroosmosis that follows applies equally to the Helmholtz and the Gouy views.

In theory, the ionic atmosphere extends an infinite distance from the surface. However, as was shown for the atmosphere surrounding a sphere by equation 4.29, the charge density, ρ, decreases more than exponentially with distance. For a flat surface, the decrease is exponential for conditions under which the Debye-Hückel approximation is valid. When **V** varies only with distance, z, from a flat charged surface, equation 4.13 can be replaced by

$$d^2\mathbf{V}/d^2z = \boldsymbol{\kappa}^2\mathbf{V} \tag{5.1}$$

By mathematical operations like those used in the first few steps of the derivation of equation 4.17 (see footnote 1 of Chapter 4), $\mathbf{V} = Ae^{-\kappa z}$, where A is the integration constant. When

$$z = 0, \mathbf{V} = \mathbf{V}_{a_i}.$$

Thus,

$$A = \mathbf{V}_{a_i} \quad \text{and} \quad \mathbf{V} = \mathbf{V}_{a_i}e^{-\kappa z} \tag{5.2}$$

As was shown in the derivation of equation 4.29,

$$\boldsymbol{\rho} = -\mathbf{D}\boldsymbol{\kappa}^2\mathbf{V}/4\pi \tag{5.3}$$

When the value of **V** from equation 5.2 is substituted,

$$\boldsymbol{\rho} = -(1/4\pi)\,\mathbf{D}\boldsymbol{\kappa}^2\,\mathbf{V}_{a_i}\,e^{-\kappa z} \tag{5.4}$$

From this it is readily apparent that if $\boldsymbol{\rho}_o$ is the value of $\boldsymbol{\rho}$ when z is zero, $\boldsymbol{\rho}/\boldsymbol{\rho}_o = e^{-\kappa z}$. Values of $\boldsymbol{\rho}/\boldsymbol{\rho}_o$ for various values of $\boldsymbol{\kappa}$z are shown in Table 5.1.

Imagine a column of unit cross section perpendicular to the charged surface extending from z = 0 to z = ∞. It will contain a number of counter charges equal but opposite in sign to the surface charge per unit area. The counter charges per unit area can be obtained by

TABLE 5.1. Values of ρ/ρ_o and of σ_3/σ for Values of κz

κz	z (cm)[a]	ρ/ρ_o	σ_z/σ	$1 - \sigma_z/\sigma$
0	0	1.000	1.000	0.000
1	0.769×10^{-7}	0.368	0.368	0.632
2	1.538×10^{-7}	0.135	0.135	0.865
3	2.308×10^{-7}	0.050	0.050	0.950
4	3.077×10^{-7}	0.0183	0.0183	0.9817
5	3.846×10^{-7}	0.0067	0.0067	0.9933
10	7.692×10^{-7}	0.000045	0.000045	0.999955

[a]for $\kappa = 1.300 \times 10^7\ \text{cm}^{-1}$.

integrating ρdz. The total counter charge per unit area between z and ∞, σ_z, is given by

$$\sigma_z = \int_z^\infty \rho dz \tag{5.5}$$

When the values of ρ from equation 5.3 and of **V** from equation 5.2 are substituted,

$$\sigma_z = \int_z^\infty (-D\kappa^2 V_{a_i}/4\pi)e^{-\kappa z}dz = -(D\kappa V_{a_i}/4\pi)e^{-\kappa z} \tag{5.6}$$

If σ is the value when $z = 0$, $\sigma_z/\sigma = e^{-\kappa z}$.

Values for various values of z are shown in Table 5.1. It can readily be calculated from its composition that Ringer's solution has an ionic strength, μ, of about 0.15. It is reasonable to use this value as an approximation of the ionic strength of most biological fluids. From equation 4.15, it can be calculated that the appropriate value of κ for biological fluids at biological temperature is 1.3×10^7 cm^{-1}. The values of z shown in Table 5.1 are calculated on the basis of this value of κ. σ_z/σ represents the fraction of the total counter charge in the atmosphere found more distant from the surface than z. $1 - \sigma_z/\sigma$, also shown in Table 5.1, is the fraction found between the surface and z. The equation and the numbers in Table 5.1 are accurate for conditions under which the Debye-Hückel approximation is valid.

It is evident from Table 5.1 that when κz has a value of 4, corresponding to a value of z of about 3 mμ, or 3 nm, 98% of the counter charges are under biological conditions no farther from a flat charged surface than z. A curved charged surface with a radius of curvature $>> 4/\kappa$ can be treated for all practical purposes as a flat surface. A circular cylinder with a charged surface containing liquid with an atmosphere of counter charges can be treated as a flat surface when the radius is greater than something like 100 times $4/\kappa$. Under biological conditions, this would be a radius of 100×3 or 300 mμ. The reason for presenting this discussion is not to provide a foundation for the development of electroosmotic theory from Debye-Hückel theory but rather to demon-

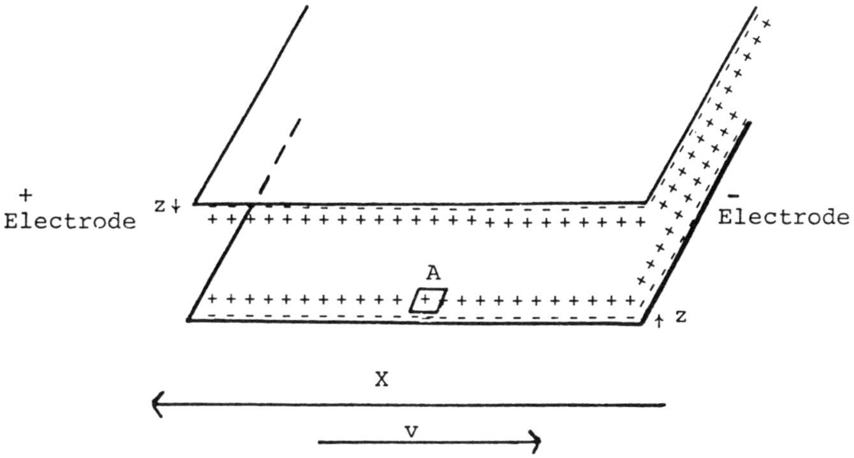

Fig. 5.1. Diagram illustrating electroosmosis. A solution of ions is enclosed between parallel nonconducting plane walls, each with uniformly distributed negative charges, with positive ionic atmospheres in the adjacent liquid. An element of volume of area, A, and thickness, dz, is located at a distance z from the bottom plane in the ionic atmosphere. An electric field of magnitude X is imposed across the liquid from right to left.

strate that most of the counter charges in the ionic atmosphere according to this widely used theory are very close to the charged surface. The development of electroosmotic theory that follows is valid for any actual distribution of counter charges in which almost all of them are close to the surface.

To develop the theory of electroosmotic flow through a capillary of radius very large compared with the region containing the counter charges, consider, as illustrated in Figure 5.1, that the liquid is enclosed between parallel nonconducting plane walls, where z is the distance from either plane. Let both plane walls have equal uniformly distributed negative charges, with positive ionic atmospheres in the adjacent liquid, very close to the walls. Apply a potential gradient, X, in the direction right to left, with the negative side on the right. In an element of volume of area, A, and thickness, dz, in the atmosphere surrounding the bottom plane, the charge enclosed will be $\rho A dz$, and the electrical force toward the right will be $X \boldsymbol{\rho} A dz$. This will cause the element of volume to move toward the right. In the steady state, the element of volume will be subjected to two additional forces, frictional forces given by Newton's law of flow, equation 2.1. On top there will be an accelerating force of

$A\eta dv/dz$ and on the bottom a retarding force of $A\eta\,[dv/dz - (d/dz)(dv/dz)dz]$. When the steady state is attained,

$$X\boldsymbol{\rho}Adz + A\eta dv/dz - A\eta[dv/dz - (d^2v/dz^2)dz] = 0.$$

This is solved for $X\boldsymbol{\rho}$ to give

$$X\boldsymbol{\rho} = -\,\eta d^2v/dz^2 \tag{5.7}$$

The potential, **V**, at the element of volume depends on two contributions, that from the charged surface and its atmosphere of counter charges, and that from the potential gradient, XL, where L is the distance in the x-direction from the positive electrode. It is apparent that $d^2\mathbf{V}/dx^2$ is zero, and, since **V** does not vary in the y-direction because the surface charge is uniform, the Poisson equation, equation 4.7, becomes

$$\nabla^2\mathbf{V} = d^2\mathbf{V}/dz^2 = -4\pi\boldsymbol{\rho}/\mathbf{D}$$

and

$$\boldsymbol{\rho} = -(\mathbf{D}/4\pi)d^2\mathbf{V}/dz^2 \tag{5.8}$$

When equation 5.8 is substituted into equation 5.7,

$$(X\mathbf{D}/4\pi)\,d^2\mathbf{V}/dz^2 = \eta d^2v/dz^2 \tag{5.9}$$

$$(X\mathbf{D}/4\pi)\,d\mathbf{V}/dz = \eta dv/dz + B_1.$$

Since $d\mathbf{V}/dz$ and dv/dz must both be zero halfway between the walls, B_1 must be zero.

$$(X\mathbf{D}/4\pi)\,d\mathbf{V}/dz = \eta dv/dz \tag{5.10}$$

$$X\mathbf{D}\mathbf{V}/4\pi\eta = v + B_2.$$

Assume no slippage at the surface. Then v = 0 when

$$\mathbf{V} = \mathbf{V}_{\mathbf{a}_i}.$$

$$v = (X\mathbf{D}/4\pi\eta)(\mathbf{V} - \mathbf{V}_{\mathbf{a}_i}) \quad (5.11)$$

When $\zeta \equiv \mathbf{V}_{\mathbf{a}_i} - \mathbf{V}_\infty$ and $u \equiv v_\infty/X$, where $\mathbf{V}_\infty$ and v_∞ are the values in the bulk of the liquid far from the surface

$$u = -\mathbf{D}\zeta/4\pi\eta \quad (5.12)$$

Equation 5.12 is the equation for electroosmotic mobility, u, as a function of the electrokinetic or zeta potential, ζ. The negative sign signifies that the liquid moves in the direction of the negative side of the applied potential gradient, X.

In some of the older literature there is a lot of discussion concerning just where shear occurs. For a smooth flat surface, equation 5.10 shows clearly that the velocity gradient is everywhere proportional to the potential gradient perpendicular to the surface. Thus shear is greatest adjacent to the surface where the potential gradient is greatest and becomes negligible a short distance from the surface where the potential gradient approaches zero. A valid objection to equation 5.12 remains, however. **D** and η are the values within the ionic atmosphere near the charged surface. To the extent that **D** and η vary with electric intensity, these are not constants, but variables. While in principle it should be possible to measure **D** and η as functions of electric field strength, the normal practice is to assume that they have the values obtained in the absence of intense electric fields. In any case, $\mathbf{D}\zeta/\eta$ is measured unambiguously by u.

Equation 5.12 as derived is actually the equation for the mobility between flat charged surfaces. If the region containing practically all of the ionic atmosphere is very small compared with the radius of a cylindrical tube of circular cross section, equation 5.12 is valid for the electroosmotic velocity in the tube.

STREAMING POTENTIAL

Streaming potential is the obverse of electroosmosis. In this case pressure causes the liquid to move past the charged nonconducting surface, carrying with it the counter ions. These build up a potential along the length of the capillary, the streaming potential, **H**. Bull [8] presented a simple derivation of the equation for streaming potential assuming that the counter ions were at a fixed distance from the charged surface, the original Helmholtz assumption.

If exactly the same assumptions are made as were made in deriving equation 5.12 for electroosmosis, the streaming potential equation can be derived for *any* distribution of counter charges. Equation 5.11 can be rewritten in the form

$$v/X = -(\mathbf{D}/4\pi\eta)\Psi, \tag{5.13}$$

where $\Psi \equiv \mathbf{V}_{\mathbf{a}_i} - \mathbf{V}$.

The velocity of a layer dz thick of liquid z cm from the charged solid surface is v and the potential at that layer is Ψ. For every element of volume, dV, within that layer, the charge, q, is ρdV. When the layer moves with a velocity, v, the counter-ion current, **I**, is $2\pi r(dz)qv$. The driving force per unit length, F′, is $2\pi r(dz)qX$.

$$\frac{2\pi r(dz)qv}{2\pi r(dz)qX} = \frac{\mathbf{I}}{F'} = \frac{v}{X} \tag{5.14}$$

Note that equation 5.14 is valid for values of dz of any thickness because dz and q cancel out. From equations 5.13 and 5.14

$$\frac{v}{X} = \frac{\mathbf{I}}{F'} = -\mathbf{D}\Psi/4\pi\eta \tag{5.15}$$

When Δz is a relatively thick layer adjacent to the surface, an average value of the potential, Ψ, must be used.

$$\Psi \equiv \mathbf{V}_{\mathbf{a}_i} - \mathbf{V} \text{ and } \zeta \equiv \mathbf{V}_{\mathbf{a}_i} - \mathbf{V}_\infty = \mathbf{V}_{\mathbf{a}_i}.$$

As was shown in equation 5.2,

$$\mathbf{V} = \mathbf{V}_{a_i}\, e^{-\kappa z}$$

Thus,

$$\Psi = \zeta(1 - e^{-\kappa z})$$

$$\overline{\Psi} = \int_o^z \Psi dz \Big/ \int_o^z dz = \left(\zeta - \zeta \int_o^z e^{-\kappa z}\right) dz \Big/ \int_o^z dz = \zeta + (\zeta/\kappa z)(e^{-\kappa z} - 1)$$

The exponential term on the right < 0.01 for $\kappa z > 5$. Thus, $\overline{\Psi} = \zeta(1 - 1/\kappa z)$. In a capillary, the largest possible value of z is r_1. If r_1 is 1 μm or 10^{-4} cm and κ, given by equation 4.15 for an aqueous solution under biological conditions at $\mu = 0.15$, is 0.13×10^8, $\kappa z = \kappa r_1$ is 1,300. Thus, for all practical purposes, when $r >> 1/\kappa$, $\overline{\Psi} = \zeta$. Therefore, for electroosmosis,

$$\mathbf{I}/F' = -\mathbf{D}\zeta/4\pi\eta \qquad (5.16)$$

In equation 5.16, **I** is the total counter-ion current, and F′ is the total force/unit length applied to the liquid in the capillary.

If equation 5.16 is valid in electroosmosis for a large capillary regardless of the distribution of counter charges, it should also be valid for streaming potential in the same capillary. The only difference is that F′ derives from pressure instead of from the interaction of charges with a potential gradient, X. For a capillary of length, L, and radius, r_1, acted on by a pressure, P,

$$F' = \pi r_1^2 P/L \qquad (5.17)$$

The stream of counter ions will build up a streaming potential, **H**, across the length of the capillary. This will cause a current to flow in the opposite direction. From Ohm's law,

$$\mathbf{I} = \mathbf{H}\Lambda_s \pi r_1^2/L \qquad (5.18)$$

Λ is the specific conductance. The significance of the subscript will be explained below.

To preserve electrical neutrality, this current, **I**, must be exactly equal to the counter-ion current of equation 5.16. When equations 5.17 and 5.18 are substituted into equation 5.16 and the result is solved for **H**,

$$\mathbf{H} = -\mathbf{DP}\zeta/4\pi\eta\Lambda_s \tag{5.19}$$

This is the original equation of Helmholtz. It is valid for capillaries in which $r_1 >> 1/\kappa$, regardless of the distribution of counter ions.

When r_1 is not very much greater than $1/\kappa$, equation 5.19 fails. Suppose κr_1 were 20. The κz could be no greater than 20, and $\overline{\Psi}$ would be $\zeta(1 - 1/20)$ or 0.95ζ. **H** is really proportional to $\overline{\Psi}$. This is the potential that would be calculated from equation 5.19; it would underestimate ζ. A second problem is encountered when κr_1 is small. Equation 5.19 was derived on the assumption that the charged solid surface is flat. Equation 5.12 and Table 5.1 show that the potential in the counter-ion atmosphere about a flat charged surface decreases to about 5% of $\mathbf{V}_{\mathbf{a}_i}$ when κz is 3. Since by equation 5.4, charge density, ρ, is proportional to $\mathbf{V}_{\mathbf{a}_i}\ e^{-\kappa z}$ this means that the greater part of the counter-ion atmosphere is no farther from the surface than $3/\kappa$. When $1/\kappa$ is 10^{-7}, as in a 0.10-μm aqueous solution at 25°C, the bulk of the atmosphere is within 3×10^{-7} cm of the surface. It is safe to use theory for a flat surface for a cylinderical surface only when $r >> 3/\kappa$.

Briggs [9] pointed out that the specific conductance, Λ, in a capillary of a porous diaphragm often differs from that of the bulk liquid. He attributed this to surface conductance. The current **I** in equation 5.18 will be proportional to the specific conductance in the capillary or porous diaphragm, Λ_s. If pure water were forced through a capillary or porous diaphragm having a high surface charge, most of the current carrying ions would be the counter ions near the surface. If, in contrast, a highly conducting solution such as 0.1 M KCl with a specific conductance, Λ, of 0.013 ohms^{-1} were forced through the same system, almost all of the current would be carried by the K^+ and Cl^- ions. Briggs' solution to the problem is to measure the system when filled with a solution of known relatively high specific conductance, Λ. Since the resistance $R = L/A\Lambda$, L/A can be calculated for the system. Next, the resistance of the capillary or porous diaphragm filled with the experimental liquid is measured. A new value of R is obtained equal to $L/A\Lambda_s$.

The value of L/A having been established in the first measurement, Λ_s is calculated, the specific conductance, including surface conductance, in the system. This is the quantity that must be used in solving equation 5.19.

When a streaming potential is generated, electroosmotic flow in the direction opposite to the streaming occurs. Bull [10] and Abramson [11] independently pointed this out. The magnitude of this effect is easily determined. The volume of liquid V that tends to flow in the opposite direction through the capillary by electroosmosis is $V = \pi r_1^2 vt$. If it is considered that this counter flow is the result of a counter pressure, P_e, by Poiseuille's law, equation 2.4, $V = \pi P_e t r_1^4/8\eta L$. When these two values of V are equated and solved for v, the electroosmotic velocity,

$$v = P_e r_1^2/8\eta L.$$

$$u = v/X = vL/\mathbf{H} = P_e r_1^2/8\eta \mathbf{H} \quad (5.20)$$

When the value for the electroosmotic mobility, u, given by equation 5.20, is substituted into equation 5.12 and solved for P_e,

$$P_e = -8H\,\mathbf{D}\zeta/4\pi r_1^2 \quad (5.21)$$

If P_s and P are the effective streaming pressure and the applied pressure, respectively, $P_s = P - P_e$. P_s is pressure that should be used to solve equation 5.19 and can be obtained by solving it. From equations 5.19 and 5.21,

$$P = P_s + P_e = -4\pi\eta\Lambda_s\mathbf{H}/\mathbf{D}\zeta - 8\mathbf{HD}\zeta/4\pi r_1^2 \quad (5.22)$$

Equations 5.19 and 5.22 can be solved to yield

$$P/P_s = 1 + (\mathbf{D}\zeta)^2/2\pi^2 r_1^2\eta\Lambda_s \quad (5.23)$$

Thus P/P_s can be greater than 1 or P_s/P less than 1 when r_1 and Λ_s are very small and ζ is very large. Whe P_s is significantly less than P, the value of ζ calculated from equation 5.19 will be less than the true value. Lauffer and Gortner [12] carried out a series of experiments on eight esters streamed through porous cellulose diaphragms. Esters have very low conductivities. Values of Λ_s for most of them were of the order of 10^{-10} or 10^{-11} mhos. In order to reduce the magnitudes of **H** to be measured, the investigators attached an 11 mohm shunt across the diaphragm. This increased the effective value of Λ_s to the order of 10^{-7} mhos. In every case the calculated value of ζ was from 5% to 50% higher for the shunted systems with the higher effective value of Λ_s than for the unshunted systems. This result is consistent with the prediction from equation 5.23.

ELECTROPHORESIS

Equation 5.12 actually expresses the relative motion of liquid and solid. When the solid surface is stationary, the equation is as written. But if the liquid is considered to be the stationary reference, then the solid surface would move in the opposite direction. That is electrophoresis. Thus, for solids that can be considered to have plane surfaces oriented parallel to X, an equation like equation 5.12 but of opposite sign applies.

$$u = \mathbf{D}\zeta/4\pi\eta \tag{5.24}$$

This would be valid for a long cylindrical charged insulator oriented parallel to X whenever the ionic atmosphere is confined to a region much smaller than its radius. Smoluchowski [13] showed that it applies to particles of any shape in any orientation, as long as the above condition applies. Experimental evidence reviewed by Abramson [2] on proteins adsorbed onto the surface of microscopic particles of many different shapes showed that this was indeed the case, as long as the particles were very large compared with $1/\kappa$.

The reasonableness of the theoretical conclusion that, when the above conditions obtain, electrophoretic mobility is independent of shape and orientation can be demonstrated by considering the situation illustrated in Figure 5.2. Assume that a spherical particle of radius r_1 has a negative charge and that r_1 is very much greater than $1/\kappa$. Under these conditions practically all of the counter ions will be very close to the surface of the sphere. Now imagine that the spheres are immobile and that the fluid streams past by electroosmosis when a potential gradient X is applied. More realistically, assume that the frame of reference moves with the spheres and that the motion of the liquid with

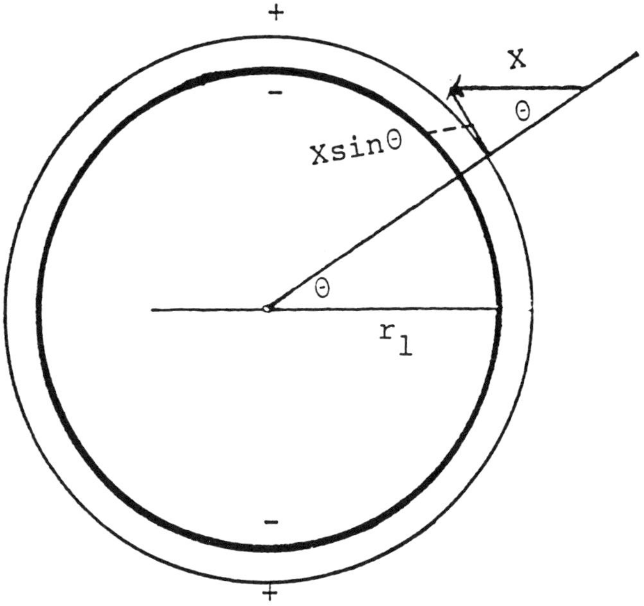

Fig. 5.2. Cross-sectional diagram of a nonconducting sphere of radius r_1 with a negative charge and a positive ionic atmosphere close to the surface immersed in an ionic solution on which an electric field of X is imposed from right to left.

respect to this frame of reference is in the opposite direction. When the sphere is a nonconductor, X will be the same everywhere in the liquid but zero in the sphere. As illustrated in Fig. 5.2 at a region very close to the surface where the counter ions are, at a point where an angle θ with the x-direction is made, the potential gradient will have a component tangential to the surface equal to X sin θ. By equation 5.12, since $u = v/X$, this will impart a tangential velocity to the counter ions, $v_\theta = -X \sin\theta\, \mathbf{D}\zeta/4\pi\eta$. But v_θ is v sin θ, where v is the velocity of the counter ions and therefore of the liquid in the x-direction. $v \sin\theta = -X \sin\theta\, \mathbf{D}\zeta/4\pi\eta$. Because all forces acting on the sphere in this case are shearing forces within the counter-ion atmosphere, this equation satisfies the condition that there be no slippage at the surface, $u = v/X = -\mathbf{D}\zeta/4\pi\eta$. This same line of reasoning will apply to particles of irregular shape, provided that at every point on the surface the radius of curvature is very large compared with $1/\kappa$. In the case of a cylinder, the line of reasoning is valid for motion perpendicular to a long axis, again providing that the radius of the cylinder is much greater than $1/\kappa$. Since it was previously shown that equation 5.24 is valid for a cylinder oriented parallel to X, electrophoretic mobility of a uniformly charged cylinder is independent of its orientation.

While equation 5.24 is valid for a large sphere in an electrolyte solution sufficiently concentrated that $1/\kappa$ is very much less than the radius, it is not valid for the opposite extreme, a small charged sphere in an extremely dilute electrolyte solution in which $1/\kappa$ approaches infinity. In this case the counter ions will be at a great distance from the center of the charged sphere. Again, if one imagines that the sphere is stationary, the countercharges will move in a potential gradient dragging the bulk of the liquid with them. Because these countercharges are very far from the surface of the stationary sphere, they will tend to move in straight lines. The situation will be comparable to that imagined when Stokes' law, equation 2.15, was derived. It will be remembered that in this case, to fulfill the condition of no slippage at the surface, pressure as well as shear forces had to be taken into account. If we revert now to a consideration of movement of the small charged sphere in the stationary liquid with counter ions at an infinite distance, two forces will be exerted on the sphere. The first is the product of the charge and the potential gradient, qX. The second is the force of frictional resistance given by Stokes' law, equation 2.15. In the steady state they are of equal magnitude. Thus, $qX = 6\pi\eta rv$. For a small isolated charged sphere, the potential at the surface, which is equal to ζ, is given by equation 4.4, $\zeta = q/\mathbf{D}r$. When q is eliminated from these two equations and the result is solved for electrophoretic mobility, one obtains

$$u = v/X = \mathbf{D}\zeta/6\pi\eta \tag{5.25}$$

Equation 5.25 was first proposed by Hückel [14]. At first there was debate whether the Hückel or the Helmholtz-Smoluchowski equation was the correct one to use to interpret electrophoretic mobility. The situation was clarified by Henry [15]. The mathematics of Henry's derivation is beyond the scope of this book, but let it suffice to say that it takes into account the path of the counter ions in their motion relative to that of the charged nonconducting spheres. Henry's solution can be expressed in the form of equation 5.26.

$$u = D\zeta f(\kappa r)/6\pi\eta \qquad (5.26)$$

In this equation, $f(\kappa r)$, read as a function of κr, varies from 3/2 when κr is much greater than 1 to 1 when κr is very much less than 1. Table 5.2 shows values of $f(\kappa r)$ for several values of κr.

What Henry actually did was to take into account the "electrophoretic" retardation effect already discussed in Chapter 4 as a factor influencing the conductance of ions. However, as was also discussed in Chapter 4, an additional effect, the time of relaxation effect, must also be considered. In the treatment of conductivity, these two effects were considered to be simply additive. Reasonable agreement with data was achieved. However, in the case of large particles, the interaction of these two effects must be taken into account. What Henry did was to re-evaluate the "electrophoretic" retardation effect by taking into account the distortion of the distribution of counter ions resulting from the time of relaxation effect. However, he ignored the direct contribution of the time of relaxation effect.

Overbeek [16] and Booth [17] carried the analysis a step further. They added to Henry's treatment a consideration of the time of relaxation effect as modified by the movement of the ions involved in the "electrophoretic" retardation effect. Their differential equation could be solved only by expansion into series. The first term is that of Henry, and additional terms involve higher powers of ζ. A more exact solution of the equation was obtained by Wiersema, et al. [18] by numerical computation with a computer. Their results are shown graphically in Figure 5.3. E is the dimensionless quantity, $6\pi\eta\epsilon u/DkT$, and y_o is $\epsilon\zeta/kT$. In the same notation, Henry's equation would be $E = y_o f(\kappa a_i)$. It can be observed by comparing with Table 5.2 that when κa_i is either very small, 0.001, or very large, 1,000, the result is the same as that given by Henry. At intermediate values of κa_i, large departures from Henry's result were obtained. This result is for a uni-univalent electrolyte and

TABLE 5.2. Henry's Function for Charged Nonconducting Spheres

κr	$f(\kappa r)$	κr	$f(\kappa r)$
0	1.000	5	1.160
1	1.027	10	1.239
2	1.066	25	1.370
3	1.101	100	1.460
4	1.133	∞	1.500

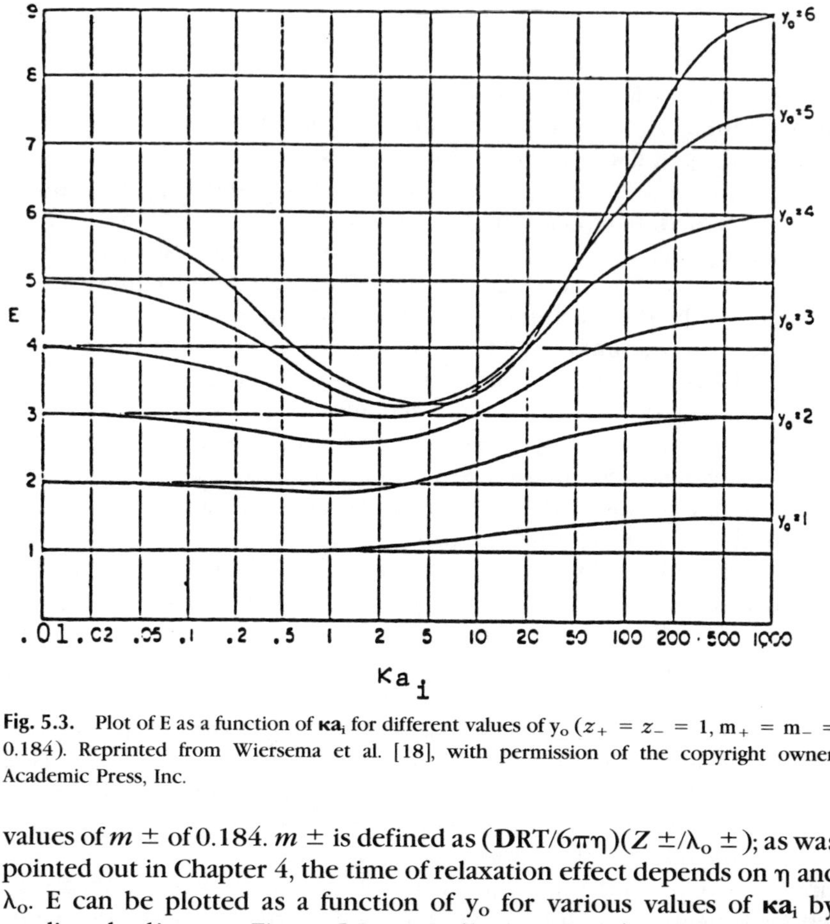

Fig. 5.3. Plot of E as a function of κa_i for different values of y_o ($z_+ = z_- = 1$, $m_+ = m_- = 0.184$). Reprinted from Wiersema et al. [18], with permission of the copyright owner, Academic Press, Inc.

values of $m \pm$ of 0.184. $m \pm$ is defined as $(DRT/6\pi\eta)(Z \pm/\lambda_o \pm)$; as was pointed out in Chapter 4, the time of relaxation effect depends on η and λ_o. E can be plotted as a function of y_o for various values of κa_i by reading the lines on Figure 5.3 vertically. Figure 5.4 is such a plot for $\kappa a_i = 5$. Curve I is the uncorrected Henry result, II is the approximation of Overbeek, III is that of Booth, and IV is the numerical computation of Wiersema et al. A more detailed discussion is provided by Overbeek and Wiersema in Bier [5], in which Figures 5.3 and 5.4 can be found.

CHARGE PER UNIT AREA OF A SPHERE

The charge per unit area of a sphere can be determined directly from the Debye-Hückel theory, equation 4.18. Since ζ is identified as V_{a_i} and the charge per unit area, σ, is given by $4\pi r^2\sigma = q$,

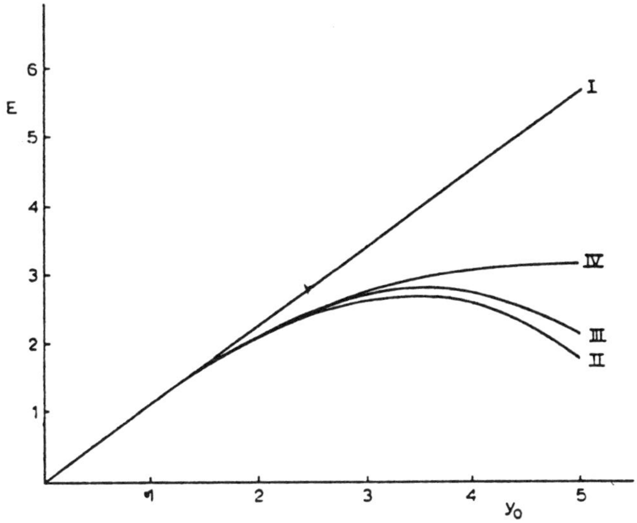

Fig. 5.4. Plot of E as a function of y_o for $z_+ = z_- = 1$, $\kappa a_i = 5$, $m_+ = m_- = 0.184$. **I:** Henry's equation. **II:** Overbeek's equation. **III:** Booth's equation. **IV:** Wiersema et al., numerical results. Reprinted from Wiersema et al. [18], with permission of the copyright owner, Academic Press, Inc.

$$\sigma = \mathbf{D}\zeta\kappa a_i^2 (1 + 1/\kappa a_i)/4\pi r^2 \qquad (5.27)$$

When κ and a_i are both sufficiently large so that $1/\kappa a_i << 1$ and $r \simeq a_i$,

$$\sigma = \mathbf{D}\zeta\kappa/4\pi \qquad (5.28)$$

It should be observed that if d is substituted for $1/\kappa$, where d is the distance between the charged plates, equation 5.28 is the equation for the relation of charge per unit area to the potential between the parallel plates in a condenser. This does not mean that $1/\kappa$ *is* the distance between the charged solid surface and the counter charges in the liquid; as shown in Table 5.1, it is the distance from the surface at which ρ is $e^{-1} = 0.37$ times the value where z is zero. Also, $1/\kappa$ is the mean distance of the counter ions from the surface.[1]

Equations 5.27 and 5.28 are valid only under conditions where the Debye-Hückel approximation is valid. When $z\epsilon V_{a_i}/kT$ is greater than 1, Table 4.2 must be used instead. The relationship between mobility and surface charge density is readily obtained, when the Debye-Hückel approximation is valid, by solving equation 5.27 for ζ and substituting into equation 5.26.

$$u = 2f(\kappa r)r^2\sigma/[3\eta\kappa a_i^2(1 + 1/\kappa a_i)]$$

$$\sigma = 3\eta\kappa a_i^2(1 + 1/\kappa a_i)\, u/2f(\kappa r)r^2 \qquad (5.29)$$

When κa_i is very large, $f(\kappa r)$ approaches 1.5 and a_i^2 approaches r^2. Equation 5.29 can then be simplified to

$$\sigma = \eta\kappa u$$

CHARGE PER UNIT AREA OF A FLAT SURFACE

When the positive and negative ions are of the same valence, $z_+ = -z_- \equiv z$, and equation 4.11 for a flat surface can be written

$$\frac{d^2V}{dz^2} = \nabla^2 V = (8\pi\nu_o \epsilon z/D)\sinh z\epsilon V/kT. \qquad (5.30)$$

Remember that z means valence and z means distance from the surface. Equation 5.31 can be derived[2] from equation 5.30.

$$\sigma = \pm\sqrt{2D\,\nu_o\,kT/\pi}\,\sinh z\epsilon\zeta/2kT \qquad (5.31)$$

From equation 4.13, the definition of the Debye-Hückel constant, κ,

$$\kappa \equiv \sqrt{4\pi\epsilon^2\Sigma\nu_i z_i^2/DkT}.$$

When the electrolyte dissociates into two ions of equal valence,

$$z,\ \Sigma\nu_i z_i^2 = 2\nu_o z^2.$$

If this substitution is made and equation 5.31 is multiplied on the right by κ and divided by the definition of κ,

$$\sigma = (D\kappa T/\pi\epsilon z)\,\kappa\,\sinh z\epsilon\zeta/2kT \qquad (5.32)$$

When $z\epsilon\zeta/DkT < 1$, $\sinh z\epsilon\zeta/2kT \simeq z\epsilon\zeta/2kT$. Thus, in the limit when

ζ is small enough, $\sigma = \mathbf{D}\zeta\kappa/4\pi$. This is the same as equation 5.28 for a large sphere when $\kappa\mathbf{a}_i >> 1$ and $r \simeq \mathbf{a}_i$.

Equation 5.32 does not involve the Debye-Hückel approximation. Thus no limit on the magnitude of $z\epsilon\zeta/2\mathbf{k}T$ is imposed. It should be valid for particles of *any* shape provided only that the radius of curvature at all points on the surface is much greater than $1/\kappa$.

CHARGE ON A CYLINDRICAL SURFACE

As was observed in previous chapters, the mathematical operations in deriving equations for cylinders are more complex than those for spheres or flat surfaces. Hill [19] has presented an equation for an infinite cylinder that can readily be transformed to equation 5.33. This is the equation for the electrokinetic potential of a cylinder of length, L, and radius, r_1, with a uniformly distributed charge on the surface of q. $\mathbf{a}_i$ has its usual meaning as the sum of r_1 and the mean radius for the counter ions. $K_0(\kappa\mathbf{a}_i)$ and $K_1(\kappa\mathbf{a}_i)$ are modified Bessel functions of the second kind, also known as modified Hankel functions, the numerical values of which are available in tabular form [20].

$$\zeta = (2q/\mathbf{D}L)\,[(K_0(\kappa\mathbf{a}_i)/\kappa\mathbf{a}_i K_i(\kappa\mathbf{a}_i) + \ln \mathbf{a}_1/r_1] \qquad (5.33)$$

Gorin has presented an equation for a long cylinder neglecting end effects identical to equation 5.33 except that L is replaced by $(L + 2r_1)$ [21]. Selected values of the Hankel functions for various values of x are shown in Table 5.3. The sixth column of that table shows that $K_0(x)/xK_1(x)$ is approximately equal to $\ln(1 + 1/x)$, with an error of less than 1% for values of x greater than 1 and less than 3% for values of x as low as 0.02. When this substitution is made and when r_1 is much greater than the mean radius of the counter ions, equation 5.33 becomes

$$\zeta = (2q/\mathbf{D}L)\ln(1 + 1/\kappa r_1) \qquad (5.34)$$

This is exactly the same as the equation for the potential, neglecting end effects, between the plates of a cylindrical condenser when L is much greater than r_1 and $1/\kappa$ is the distance between the plates.[3]

An alternative to equation 5.33 that does attempt to take into account end effects was derived by Lauffer [22]. Advantage is taken of

TABLE 5.3. Modified Bessel Functions of the Second Kind (Hankel Functions)

(x)	$2/\pi K_o(x)$	$2/\pi K_1(x)$	$K_o(x)/xK_1(x)$	$\ln(1 + 1/x)$	$\frac{[K_o x/xK_1(x)]}{\ln(1 + 1/x)}$	1/x
0.02	2.565	31.802	4.0328	3.9318	1.026	50.0
0.04	2.124	15.867	3.3466	3.2581	1.027	25.0
0.06	1.867	10.545	2.9508	2.8717	1.028	16.667
0.08	1.685	7.878	2.6736	2.6027	1.027	12.500
0.10	1.5451	6.273	2.4631	2.3979	1.027	10.000
0.14	1.3351	4.432	2.1517	2.0971	1.026	7.1429
0.20	1.1158	3.0405	1.8349	1.7918	1.024	5.0000
0.30	0.8737	1.9455	1.4970	1.4663	1.021	3.3333
0.50	0.5885	1.0545	1.1162	1.0986	1.016	2.0000
0.70	0.4205	0.6686	0.8985	0.8873	1.013	1.4286
1.00	0.2680	0.3832	0.6994	0.6931	1.0091	1.000
1.40	0.15512	0.2043	0.5423	0.5390	1.0062	0.7143
2.0	0.07251	0.08904	0.4072	0.4055	1.0042	0.5000
3.0	0.02212	0.02556	0.2885	0.2877	1.0027	0.3333
4.0	0.007104	0.007947	0.2235	0.2231	1.0015	0.2500
5.0	0.002350	0.002575	0.1825	0.1823	1.0011	0.2000
6.0	0.0007920	0.0008556	0.1543	0.1542	1.0008	0.1667
10.0	0.11319×10^{-4}	0.11872×10^{-4}	0.09534	0.09531	1.0003	0.1000
15.0	0.6251×10^{-7}	0.6456×10^{-7}	0.06455	0.06454	1.0002	0.0667

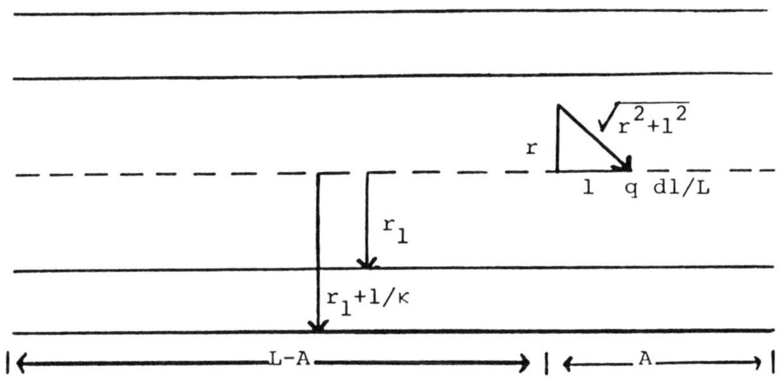

Fig. 5.5. Cross-sectional diagram of a cylinder of length L and radius r_1 and uniformly distributed charge q (which acts as though it were a uniform line charge on the axis of the cylinder) lying within a concentric cylinder in the liquid of radius $r_1 + 1/\kappa$.

the parallelism noted in equation 5.34 and previously in equation 5.28 of the identity of electrokinetic equations with those for condensers in which the distance between the plates is substituted for $1/\kappa$. The model is a cylinder of length, L, and radius, r_1, with a uniformly distributed fixed charge, q, on the cylindrical surface, separated from a concentric cylinder with a radius of $r_1 + 1/\kappa$. Symmetry permits this to be treated as a problem of a uniform line charge, q/L, located at the axis of the cylinders. The problem can be solved by reference to Figure 5.5. By equation 4.2, the scalar magnitude of the electric intensity at the end of r resulting from the charge q dl/L is $(q\ dl/L)/D(r^2 + l^2)$. The magnitude of its component in the direction of a radius, E_{Adl}, is this multiplied by $r/\sqrt{r^2 + l^2}$. Thus the component in the radial direction, E_{Adl}, at the end of a cylindrical radius, r, originating at a point A distant from the end of the cylinder on the line arising from an element, q dl/L, of charge located l cm from A, is

$$E_{Adl} = q\ dl\ r/DL(r^2 + l^2)^{3/2}$$

Consider the point, A, as the origin. The total electric intensity in the radial direction at r, E_A, is obtained by integrating between $-A$ and $L - A$

$$E_A = qr/DL) \int_{-A}^{L-A} dl/(r^2 + 1^2)^{3/2} =$$

$$(q/DrL) \left[(L - A)/(r^2 + (L - A)^2)^{1/2} + A/(r^2 + A^2)^{1/2} \right]$$

If ζ_A is the potential difference between two points at $r_1 + 1/\kappa$ and r_1 on the radius originating at A,

$$\zeta_A = \int_{r_1}^{r_1 + 1/\kappa} E_A \, dr$$

When the value for E_A above is substituted, the resultant equation can be integrated exactly. However, the result is cumbersome. When κr_1 is much greater than 1, an approximation sufficiently accurate for most purposes can be obtained by integrating the following equation, where the quantity outside the integral sign is treated as a constant.

$$\zeta_A = (q/DL) \left[(L - A)/(r^2 + (L - A)^2)^{1/2} + A/(r^2 + A^2)^{1/2} \right] \int_{r1}^{r_1 + 1/\kappa} dr/r$$

The approximation involves substituting outside the integral $r_1 + 1/2\kappa$ for r, the mean value of r over the range of the integration. The skeptical reader can ascertain by numerical computations that this approximation leads to accurate results over a wide range of dimensions.[4] The result of this approximation is a constant multiplied by the integral of dr/r and has the value

$$\zeta_A = q/DL \left[\frac{L - A}{[(r_1 + 1/2\kappa)^2 + (L - A)^2]^{1/2}} + \frac{A}{[(r_1 + 1/2\kappa)^2 + A^2]^{1/2}} \right] \ln (1 + 1/\kappa r_1)$$

This equation shows that the potential difference, ζ_A, is not a constant but varies with the position A. The average value,

$$\zeta, \text{ is } \int_0^L \zeta_A dA \Big/ \int_0^L dA$$

When the equation for ζ_A is substituted and the integration carried out, equation 5.35 results. Unfortunately, there is a printing error in the original publication [19] of this equation.

$$\zeta = (2q/DL^2)\ln(1 + 1/\kappa r_1)\left[[(r_1 + 1/2\kappa)^2 + L^2]^{1/2} - (r_1 + 1/2\kappa)\right] \quad (5.35)$$

When L is much greater than $r_1 + 1/2\kappa$, equation 5.35 also reduces to equation 5.34. Equation 5.35 has the advantage of taking into account, at least partially, the fact that the potential at the end of the cylinder is less than that far from the end. This is useful in attempting to estimate the electrical work involved in a reaction in which two short cylinders of equal charge density react to form a long cylinder [19]. While equation 5.35 takes into account end effects in a cylindrical condenser in which the cylindrical plates are separated by a distance $1/\kappa$, it probably does not estimate as accurately the end effects of a charged cylinder surrounded by counter ions in solution. The reason is that the counter ions at the ends need not all be distributed along radii of the cylinder.

When κr_1 is much greater than 1, $\ln(1 + 1/\kappa r_1)$ approaches $1/\kappa r_1$. Since for a cylinder $\sigma = q/2\pi r_1 L$, equation 5.35 reduces to equation 5.28, when $L >> r_1$. Thus, for large values of κr_1 and long cylinders, the relationship between surface charge density, σ, and electrokinetic potential, ζ, is the same for spherical surfaces, cylindrical surfaces, and plane surfaces.

Since cylinders obey equation 5.24 when $r_1 >> 1/\kappa$, equation 5.28, the simplified version of equation 5.35, and equation 5.24 can be combined to give

$$\sigma = \eta \kappa u \quad (5.36)$$

Equation 5.36 is the same as the simplified version of equation 5.29 for spheres when κa_i is very large.

At the opposite extreme, when $r_1 >> 1/\kappa$ the situation is similar to that encountered with spheres under the same condition. The electrical force is $qx = 2\pi r_1 L\sigma x$, and the force of resistance is fv. When these are equated and solved for σ, and when v/x is replaced by u, $\sigma = uf/2\pi r_1 L$.

The friction coefficient, f, of randomly oriented cylinders when $L >> r_1$ is approximately the same as that for randomly oriented elongated ellipsoids of revolution, given by the simplified form of equation 2.18 if r_1 is substituted for a and L/2 for b. Thus,

$$f \simeq 3\pi\eta L/\ln L/r_1$$

and

$$\sigma = (3/2)\ \eta u/r_1 \ln (L/r_1) \tag{5.37}$$

This should be compared with the equation for a sphere given by equation 5.29, for the case that

$$1/\kappa a_i >> 1 \text{ and } f(\kappa r) \text{ is } 1;\ \sigma = (3/2)\eta u\ a_i/r_1^2.$$

ATTEMPTS TO VERIFY A THEORY

The ultimate basis for the acceptance of any theory must be experimental verification. Many investigators have attempted to verify electrokinetic theory through studies on proteins. From many points of view, proteins would seem to be the ideal system for such a verification. They are nonconducting particles. The known sources of charge are the dissociation of acidic and basic groups and the adsorption of specific ions. In principle, it is possible to determine the contribution of dissociation from acid and base titration studies, and methods are available for measuring adsorption of specific ions. For proteins that can be considered to be spheres, the electrokinetic potential, ζ, can be calculated from u with Henry's equation, equation 5.26, or from the extensions of Henry's theory worked out by Booth and by Overbeek et al. When the electrokinetic potential, ζ, is small enough for the Debye-Hückel approximation to be valid, the surface charge density, or the number of charges per unit area of the surface, σ, can be evaluated from ζ with equation 5.27. When this is combined with equation 5.26 to get equation 5.29, σ can be calculated directly from the electrophoretic mobility, u. For very large values of the radius and of the Debye-Hückel

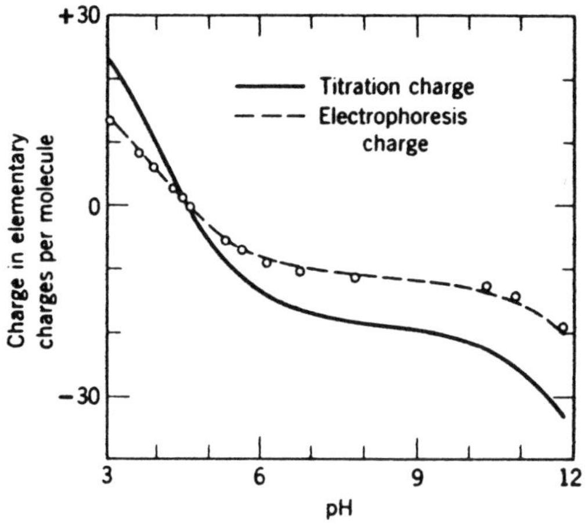

Fig. 5.6. The charge of ovalbumin as a function of pH as determined from electrophoretic mobilities (dashed curve) compared with charge determined by acid-based titration (solid curve). Reprinted from Overbeek [24], with the permission of the copyright owner, John Wiley & Sons (Interscience).

constant, **κ**, equation 5.32 can be used to calculate σ from ζ even when ζ is too large to permit validity of the Debye-Hückel approximation.

Many investigators have analyzed relevant data obtained with proteins with generally disappointing results. One of the earliest was Longsworth [23]. His data were utilized by Overbeek [24] to obtain the results shown in Figure 5.6. The unsatisfactory agreement between values of σ determined from electrophoresis and those determined from titration has not been improved substantially in any subsequent report discovered by the author. There are two possibilities. One is that the theory is wrong and the other is that proteins don't fit the models for which theory was derived, smooth spheres or cylinders with uniformly distributed surface charges.

Tanford and Kirkwood [25] developed a theory based on the contribution of isolated charges distributed across the surface in a nonuniform pattern. Lee and Richards [26] attempted to evaluate the static accessibility of the ionizing groups on the surface of the protein from crystallographic data. The electrostatic contribution of the ionizable group should depend upon whether it is totally exposed to the solvent or partially buried within the protein structure. Various attempts have been made to test the Tanford-Kirkwood formalism by the method of hydrogen ion titration. This is in a very real sense the equivalent of a test of equation 5.27 by electrophoresis measurements, because Linderstrøm-Lang [27] had shown that equation 5.27 could be used to help to understand the difference between the titration curve of a protein and that of a solution containing a mixture of the same titratible groups in the same proportion as found in the protein.[5] In one sense, considerable success was achieved in these studies [26,28,29], but the static solvent accessibility factors required to make the data fit the theory must be assigned arbitrary values, values that sometimes differ from the best estimates from crystallographic data. Thus, this much more detailed theoretical analysis doesn't work very well either. If it doesn't work for hydrogen ion titration studies, it shouldn't work for electrophoretic studies.

Even though we are denied complete assurance that electrophoretic theory is valid, it is successful in explaining some aspects of the relationship between mobility and charge. Many investigations have shown that in the region near the isoelectric point, where charge is not too high, there is direct proportionality between electrophoretic mobility and charge, even though the proportionality constant differs from that predicted by theory, as in equation 5.29. This is illustrated by the data of Schlessinger [30], shown in Figure 5.7. For charges between about +20 and −20 proton units, there is accurate proportionality. Marked deviations are apparent for large values of charge, either positive or negative. One interpretation, advocated by several investigators, is that at pH values far removed from the isoelectric point, the protein molecule changes shape so that the frictional resistance dependent upon the radius changes.

Electrophoretic theory is also successful in explaining the nonspecific effect of ionic strength on mobility. Bull [8] carried out a novel test of electrophoretic equations. He compared charge per unit surface area for dissolved bovine serum albumin, σ_D, measured by the moving boundary method at various pH values and an ionic strength of 0.05 by Schlessinger [30] with the charge per unit area of the same protein under the same conditions adsorbed onto inert particles of octadecane, σ_A. Mobilities in the latter case were measured by the microelectrophoresis method. Henry's equation, 5.26, was used to evaluate ζ from mobility in both cases. Then, for the dissolved protein, equation 5.27, the equation derived from the Debye-Hückel theory involving the correction $(1 + 1/\kappa a_1)$ was used to evaluate σ_D from ζ. For the protein adsorbed onto the large octadecane particles, Bull used equation 5.32 to calculate σ_A from ζ. This equation does not contain $(1 + 1/\kappa a_2)$. In Figure 5.8, σ_A is

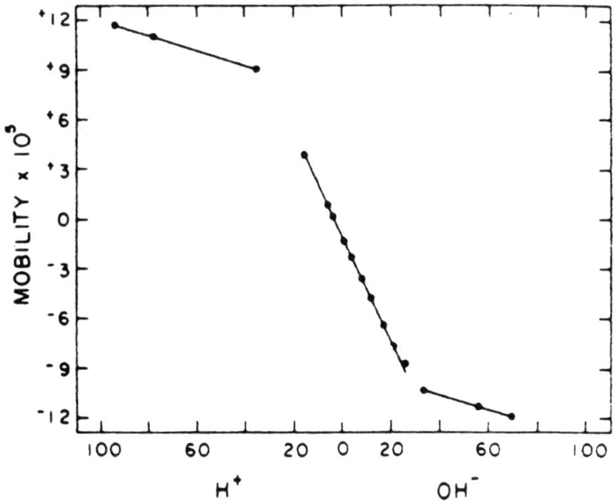

Fig. 5.7. Electrophoretic mobilities of bovine serum albumin compared with equivalents of H^+ bound per mole (ionic strength 0.01). Reprinted from Schlessinger [30], with permission of the copyright owner, The American Chemical Society.

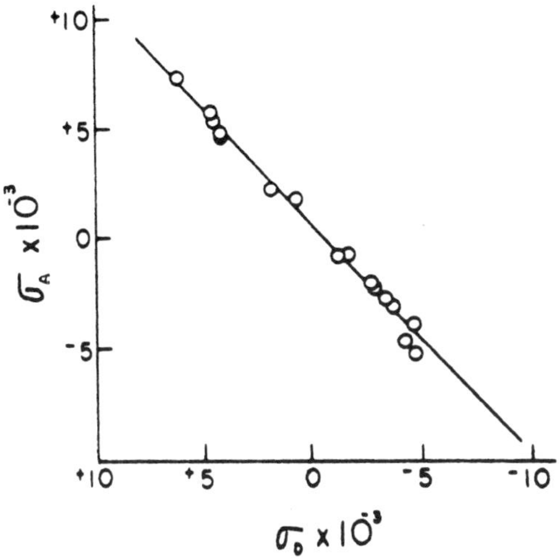

Fig. 5.8. Comparison of electrostatic charge per unit area for adsorbed (σ_A) and for dissolved (σ_D) bovine serum albumin calculated from electrophoretic mobilities. Reprinted from Bull [8], with permission of the copyright owner, FA Davis Co.

plotted against σ_D. The data fit the equation, $\sigma_A \times 10^{-3} = 0.8 \times 10^{-3} + \sigma_D \times 10^{-3}$. The constant, 0.8×10^{-3}, derives from the fact that the isoelectric points of the dissolved and of the adsorbed protein are not identical. Bull suggests that this might be the result of exposure of some positive charges when the protein is adsorbed onto the inert hydrocarbon surface. The data, however, do show that increments in charge per unit area are the same for the protein molecules that are small enough to require the Debye-Hückel correction and for the protein adsorbed onto particles large enough to be independent of that correction. This is at least suggestive evidence that equation 5.27, which involves the nonspecific ionic strength (κ) contribution, is valid.

REFERENCES

1. Freundlich, H. Colloid and Capillary Chemistry (translated into English by H. S. Hatfield). London: Methuen, 1926, p. 239.
2. Abramson, H. A. Electrokinetic Phenomena. Chemical Catalog Co., Inc.. New York: Reinhold Publishing Corp., 1934.
3. Abramson, H. A., Moyer, L. S., and Gorin, M. H. Electrophoresis of Proteins. New York: Reinhold Publishing Corp., 1942.
4. Bier, M. (ed.) Electrophoresis, Vol. I. New York: Academic Press, 1959.
5. Bier, M. (ed.) Electrophoresis, Vol. II. New York: Academic Press, 1967.
6. Helmholtz, H. Wied. Ann. *7:*337, 1879.
7. Gouy, L. J. Phys. *9:*457, 1910.
8. Bull, H. B. An Introduction to Physical Biochemistry, Second Edition, Philadelphia: F. A. Davis Co., 1971.
9. Briggs, D. J. Phys. Chem. *32:*641, 1928.
10. Bull, H. B. Kolloid Z. *60:*130, 1932.
11. Abramson, H. A. J. Gen. Physiol. *15:*279, 1932.
12. Lauffer, M. A. and Gortner, R. A. J. Phys. Chem. *43:*721, 1939.
13. v. Smoluchowski, M. In: Handbuch der Elektricität und des Magnitismus, Graetz (ed.). Leipzig: Barth, 1921. p. 366.
14. Hückel, E. Phys. Z. *25:*204, 1924.
15. Henry, D. C. Proc. R. Soc. Lond. [A] *133:*106, 1931.
16. Overbeek, J.Th.G. Kolloidchem. Beih. *54:*287, 1943.
17. Booth, F. Proc. R. Soc. Lond. [A] *203:*514, 1950.
18. Wiersema, P.H., Loeb, A.L., and Overbeek, J.Th.G. J. Coll. Intf. Sci. *22:*78, 1966.
19. Hill, T. L. Arch. Biochem. Biophys. *57:*229, 1955.
20. Lösch, F. Jahnke-Emde-Lösch Tables of Higher Functions, Sixth edition. New York: McGraw-Hill, 1960.
21. Abramson, H. A., Gorin, M. H., and Moyer, L. S. Chem. Rev. *24:*345–366, 1939.
22. Lauffer, M. A. In: Subunits in Biological Systems, S.N. Timasheff and G.D. Fasman (eds.). New York: Marcel Dekker, 1971 pp. 149–199.
23. Longworth, L. G. Ann. N.Y. Acad. Sci. *41:*267, 1941.
24. Overbeek, J.Th.G. Adv. Colloid Sci. *3:*97, 1950.

25. Tanford, C. and Kirkwood, J.G. J. Am. Chem. Soc. *79:*5333, 1957.
26. Lee, B.K. and Richards, F.M. J. Mol. Biol. *55:*379, 1971.
27. Linderstrøm-Lang, K. C.R. Trav. Lab. Carlsberg *15:*1, 1924.
28. Shire, S.J., Hanania, G.I.H., and Gurd, F.R.N. Biochemistry *13:* 2967, 1974.
29. Tanford, C. and Roxby, R. Biochemistry *11:*2192, 1972.
30. Schlessinger, B.S. J. Phys. Chem. *62:*916, 1958.

NOTES

1.

$$\bar{z} = \int_0^\infty \boldsymbol{\rho}\, z\, dz \Big/ \int_0^\infty \boldsymbol{\rho}\, dz$$

When equation 5.4 is substituted for $\boldsymbol{\rho}$,

$$\bar{z} = \int_0^\infty e^{-\boldsymbol{\kappa} z}\, z\, dz \Big/ \int_0^\infty e^{-\boldsymbol{\kappa} z}\, dz$$

$$= (1/\boldsymbol{\kappa}^2)\, e^{-\boldsymbol{\kappa} z}\, (-\boldsymbol{\kappa} z - 1) \Big|_0^\infty \Big/ (-1/\boldsymbol{\kappa})\, e^{-\boldsymbol{\kappa} z} \Big|_0^\infty = 1/\boldsymbol{\kappa}$$

2. If both sides are multiplied by $2d\mathbf{V}/dz$

$$(2d\mathbf{V}/dz)d(d\mathbf{V}/dz)/dz = (16\pi\nu_o \epsilon z \boldsymbol{\kappa} T/\epsilon z \mathbf{D}) \sinh z\epsilon\mathbf{V}/\mathbf{k}T d(z\epsilon\mathbf{V}/\mathbf{k}T)/dz.$$

Upon integration,

$$(d\mathbf{V}/dz)^2 = (16\pi\nu_o \mathbf{k}T/\mathbf{D})(\cosh z\epsilon\mathbf{V}/\mathbf{k}T) + \text{const.}$$

When $z = \infty$, $d\mathbf{V}/dz = 0$, $\mathbf{V} = 0$ and $\cosh \mathbf{V} = 1$.

$$0 = 16\pi\nu_o \mathbf{k}T/\mathbf{D} + \text{const.}$$

$$(d\mathbf{V}/dz)^2 = (16\pi\nu_o \mathbf{k}T/\mathbf{D})(\cosh z\epsilon\mathbf{V}/\mathbf{k}T - 1) = (32\pi\nu_o \mathbf{k}T/\mathbf{D}) \sinh^2 z\epsilon\mathbf{V}/2\mathbf{k}T \qquad \text{(Eq. 5.30a)}$$

The transformation of the cosh function into the $\sinh^2$ function is explained as follows:

$$2(\cosh x - 1) = e^x + e^{-x} - 2$$
$$(2 \sinh x/2)^2 = (e^{x/2} - e^{-x/2})^2 = e^x + e^{-x} - 2 = 2(\cosh x - 1)$$
$$\cosh x - 1 = 2 \sinh^2 x/2$$

From equation 4.7, $\rho = -(\mathbf{D}/4\pi)d^2V/dz^2$. When this value of ρ is substituted into equation 5.5,

$$\sigma_z = -(\mathbf{D}/4\pi)\int_z^\infty (d(dV/dz)/dz)dz = (\mathbf{D}/4\pi)\, dV/dz$$

because when z is ∞, dV/dz is zero. When this equation is solved for dV/dz and squared, $16\pi^2 \sigma_z^2/\mathbf{D}^2 = (dV/dz)^2$. When this value of $(dV/dz)^2$ is substituted into equation 5.30a, $16\pi^2\sigma_z^2/\mathbf{D}^2 = (32\pi\nu_o\mathbf{k}T/\mathbf{D})(\sinh^2 z\epsilon V/2\,kT)$. $\sigma_z^2 = (2\mathbf{D}\nu_0\mathbf{k}T/\pi)(\sinh^2 z\epsilon\, V/2\mathbf{k}T)$. At the surface, $V = \zeta$ and the surface charge per unit area is σ. When these substitutions are made and roots are extracted, equation 5.31 is obtained.

3. From equation 4.1, Gauss' law,

$$\oint \bar{E} \cdot d\bar{Q} = 4\pi q/\mathbf{D}$$
$$E\, 2\pi rL \cos 0 = 4\pi q/\mathbf{D}$$
$$E = 2q/L\mathbf{D}r$$

$$V = -\int_{r_1}^{r_2} \bar{E} \cdot dl = -\int_{r_1}^{r_2} E \cos 180\, dr =$$

$$\int_{r_1}^{r_2} (2q/L\mathbf{D})\, dr/r = (2q/L\mathbf{D}) \ln (1 + d/r_1)$$

where $d \equiv r_2 - r_1$.

This derivation of the equation for the potential across the gap in a cylindrical condenser, neglecting end effects, is found in some textbooks of general physics.

4. The differential equation is the sum of two terms that differ only in that one contains A and the other L − A. The accuracy of the approximation can be assessed by comparing the exact integral of the term involving A with the approximate integral,

$$\Psi = (qA/DL) \int_{r_1}^{r_1 + 1/\kappa} dr/r\sqrt{r^2 + A^2}$$

The exact integral is

$$\Psi_e = -(q/DL) \ln (A + \sqrt{r^2 + A^2}/r) \Big|_{r_1}^{r_1 + 1/\kappa} =$$

$$q/DL \ln \left[\left(\frac{1 + \sqrt{1 + r_1^2/A^2}}{1 + \sqrt{1 + (r_1 + 1/\kappa)^2/A^2}} \right) (1 + 1/\kappa r_1) \right]$$

The approximate integral is

$$\Psi_a = (qA/DL)\sqrt{(r_1 + 1/2\kappa)^2 + A^2} \ln(1 + 1/\kappa r_1)$$

As $A \rightarrow 0$

$$\Psi_e \rightarrow (q/DL) \ln (r_1/(r_1 + 1/\kappa))(1 + 1/\kappa r_1) = 0$$

$$\Psi_a \rightarrow 0$$

Both the exact and the approximate solutions show that Ψ approaches 0 at the ends to the cylinder. This is very different from the case of a cylindrical condenser, where Ψ is constant.

Numerical substitution shows that when $\kappa r_1 = 10$, $\Psi_a = \Psi_e$ to a very good approximation for values of A less than r_1. However, when κr_1 is substantially less than 10, the approximate integral is significantly in error over this range of A, that is, in the central portion of the cylinder. When $A >> r_1 + 1/\kappa$, both Ψ_a and Ψ_e approach $(q/DL) \ln (1 + 1/\kappa r_1)$, the value corresponding to that of a cylindrical condenser.

5. If n_o is the number of sites on a protein molecule capable of binding a proton and if n is the number protonated, $pH = pK + \log (n_o - n)/n$, where pK is $-\log K$, the dissociation constant. Assume that ΔG^o for the dissociation of a proton by a molecule is the sum of two terms, intrinsic free energy, ΔG_i^o, and an electrical contribution, ΔG_e^o, equal to $\epsilon\zeta$. Since, per molecule, $\Delta G^o = -\mathbf{k}T \ln K = 2.303\ \mathbf{k}T\ pK$,

$pK = pK_i + \epsilon\zeta/2.303\ \mathbf{k}T$. The charge on the protein molecule is $z\epsilon = 4\pi r_1^2\sigma$. Thus, equation 5.27 can be rearranged to yield $\zeta = z\epsilon/Da_i^2(\kappa a_i + 1)$. By combining all of these equations,

$$pH - \log \frac{n_o - n}{n} = pK_i + z\epsilon^2/2.303\ \mathbf{k}TDa_i^2\ (\kappa a_i + 1).$$

This equation has been partially successful in explaining the titration curves of many proteins. While the more detailed theory of Tanford and Kirkwood, as modified to include the concept of static solvent accessibility, does better, it is still not fully satisfactory because arbitrary or empirical parameters must be assigned.

6

Potentials at Interfaces

A cell's membrane creates two interfaces, one between its inner surface and the cell constituents and one between its outer surface and the external medium. In the living cell there is a potential difference between the inside and the outside, usually of a magnitude somewhere between 20 and 100 mv, with the inside being negative. Electrophysiology deals with resting potentials, injury potentials, action potentials, etc. It is the purpose of the present chapter to treat the physics and physical chemistry underlying potentials at interfaces and therefore potentials across membranes. No attempt is made to review the vast amount of information constituting the field of electrophysiology. The chapter does deal with the physics and physical chemistry of electrolytic cells, liquid junctions, potentials across membranes with selective permeability, and the like [1].

ELECTRODE POTENTIALS

A Daniell cell consists of a bar of zinc immersed in a solution of $ZnSO_4$ at molality, m_1, and a bar of copper immersed in a solution of $CuSO_4$ at molality, m_2, with an electrical contact such as a porous plug or a KCl bridge connecting the two solutions. If a conductor connects the zinc and the copper bars, when m_1 and m_2 do not differ greatly, a negative current will flow through the conductor from the zinc to the copper and a positive current from the zinc to the copper through the solutions. At the zinc electrode, electrons will be removed from zinc atoms to form Zn^{2+} ions, oxidation, and at the copper electrode, electrons will be added to Cu^{2+} ions to form copper atoms, reduction. The reaction at the zinc electrode can be represented by $Zn \rightarrow Zn^{2+}$ +

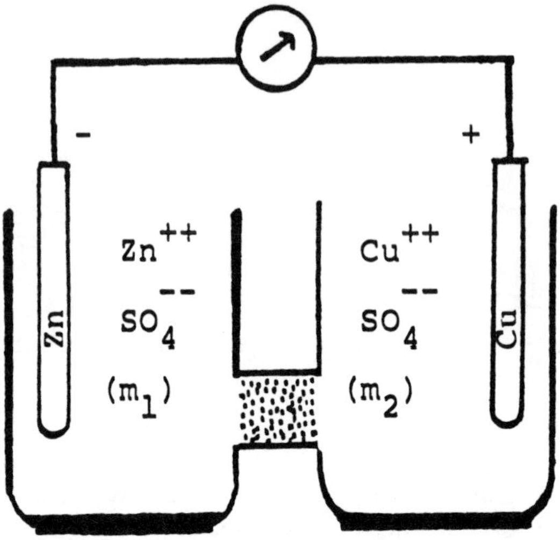

Fig. 6.1. Schematic representation of Daniell cell.

2**e**, where **e** means an electron. At the copper electrode, the reaction is $Cu^{2+} + 2\mathbf{e} \rightarrow Cu$. The net chemical reaction is $Zn + Cu^{2+} \rightarrow Zn^{2+} +$ Cu. The cell can be represented as follows:

$$Zn \mid Zn^{2+}\ (m_1) \mid\mid Cu^{2+}\ (m_2) \mid Cu \qquad \text{(cell A)}$$

If instead of the metallic conductor between the zinc and the copper electrodes, a potentiometer is attached, as illustrated in Fig. 6.1, the zinc electrode will exhibit a negative potential and the copper electrode a positive potential. The potentiometer will measure the difference between the potential on the copper electrode and that on the zinc electrode. When the cell is represented as above, with the zinc on the left and the copper on the right, and when the potentiometer is adjusted so that no current flows through it, that is, when its potential exactly balances that of the cell, the difference between the potential on the right, $\mathbf{V}_r$, and that on the left, $\mathbf{V}_l$, is equal to the electromotive force, **E**, of the cell. When the concentrations of both ions correspond to unit activity, $\mathbf{V}_r - \mathbf{V}_l$ is the standard electromotive force of the cell, $\mathbf{E}^0$. When the activities of either or both of the ions differ from unity, **E** is given by $\mathbf{E} = \mathbf{E}^0 - (RT/n\mathbf{F})\ln a_{Zn^{2+}}/a_{Cu^{2+}}$. This is the appropriate form of the Nernst equation. Its most general form for the reaction aA + bB $\rightleftharpoons$ cC + dD with the transfer of n electrons is[1]

$$\mathbf{E} = \mathbf{E}^0 - (\mathrm{RT/n}\mathbf{F})\ln(a_C^c\, a_D^d / a_A^a\, a_B^b) \qquad (6.1)$$

In the Daniell cell the number of faradays of electricity involved, n, is two.

Electromotive force is a measure of the tendency to drive positive electricity through the cell. When the Daniell cell is written with the zinc electrode on the left, as represented in Cell A, positive electricity flows within the cell from left to right. Therefore, **E** for the cell is positive. Yet the zinc electrode is negative on the outside and the copper electrode is positive on the outside. The difference between these outside potentials is the only quantity that can be measured with a voltmeter or a potentiometer.

$$\mathbf{E} = \mathbf{V}_r - \mathbf{V}_l \qquad (6.2)$$

While it is impossible to measure the potential of a single electrode, it is nevertheless convenient to divide cell potentials and standard potentials into two portions, $\mathbf{V}_r^0$ and $\mathbf{V}_l^0$

$$\mathbf{E}^0 = \mathbf{V}_r^0 - \mathbf{V}_l^0$$

When a hydrogen electrode (platinized platinum saturated with H_2 at a pressure of 1 atmosphere) is substituted on the left for the zinc electrode in Figure 6.1, a different value of $\mathbf{E}^0$ is obtained. If $\mathbf{V}^0$ for the hydrogen electrode is arbitrarily assigned a value of 0, then the value of $\mathbf{E}^0$ for this cell can be called $\mathbf{V}^0$ for the copper electrode. In this manner, except for a complication involving liquid junction potentials, discussed later, the standard reduction potentials for many electrodes can be obtained and can be arranged in an electromotive series as illustrated in Table 6.1, the most negative on top and the most positive on the bottom. Older tables listed standard $\mathbf{E}^0$ values, the same as standard reduction potentials multiplied by (-1). One can always obtain $\mathbf{E}^0$ for a cell, the important quantity in electrochemistry, by subtracting the standard reduction potential of the electrode where oxidation takes place from that of the electrode at which reduction takes place.

TABLE 6.1. Standard Reduction Electrode Potentials at 25°C

Electrode	Electrode reaction	V° (volts)
$Li^+|Li$	$Li^+ + e \rightleftarrows Li$	−3.045
$K^+|K$	$K^+ + e \rightleftarrows K$	−2.925
$Cs^+|Cs$	$Cs^+ + e \rightleftarrows Cs$	−2.923
$Ca^{2+}|Ca$	$Ca^{2+} + 2e \rightleftarrows Ca$	−2.866
$Na^+|Na$	$Na^+ + e \rightleftarrows Na$	−2.714
$Zn^{2+}|Zn$	$Zn^{2+} + 2e \rightleftarrows Zn$	−0.763
$Fe^{2+}|Fe$	$Fe^{2+} + 2e \rightleftarrows Fe$	−0.440
$Sn^{2+}|Sn$	$Sn^{2+} + 2e \rightleftarrows Sn$	−0.136
$Pb^{2+}|Pb$	$Pb^{2+} + 2e \rightleftarrows Pb$	−0.126
$Fe^{3+}|Fe$	$Fe^{3+} + 3e \rightleftarrows Fe$	−0.036
$H^+|H_2|Pt$	$2H^+ + 2e \rightleftarrows H_2$	0.000
$Cu^{2+}|Cu$	$Cu^{2+} + 2e \rightleftarrows Cu$	+0.337
$I^-|I_2|Pt$	$I_2 + 2e \rightleftarrows 2I^-$	+0.535
$Ag^+|Ag$	$Ag^+ + e \rightleftarrows Ag$	+0.799
$Hg^{2+}|Hg$	$Hg^{2+} + 2e \rightleftarrows Hg$	+0.854
$Br^-|Br_2|Pt$	$Br_2 + 2e \rightleftarrows 2Br^-$	+1.065
$Cl^-|Cl_2|Pt$	$Cl_2 + 2e \rightleftarrows 2Cl^-$	+1.359
$OH^-|H_2|Pt$[a]	$2H_2O + 2e \rightleftarrows 2OH^- + H_2$	−0.828

[a]Alkaline solution.

The zinc and copper electrodes are reversible electrodes. If the potential from the potentiometer opposing the flow through electrical connectors between the electrodes is slightly less than sufficient to prevent flow, the chemical reactions at the two electrodes will be as indicated by the chemical equation previously written. If the potential on the potentiometer is slightly greater than the amount necessary to prevent the flow of current, the chemical reaction will be the exact reverse. In contrast to the reversible character of the charge-carrying processes at the metallic electrodes, the transport of charge across a liquid junction is irreversible. When positive electricity flows from the zinc to the copper electrode, it is carried across the junction between the $ZnSO_4$ and the $CuSO_4$ solutions by the migration of Zn^{2+} ions in the forward direction and of $SO_4{}^{2-}$ ions in the reverse direction. When the current is carried in the opposite direction across the junction, Cu^{2+} and $SO_4{}^{2-}$ ions are involved. Thus liquid junctions are not reversible. Potentials, which are discussed in the next section, exist at liquid junctions. To obtain accurate data for standard electrode reduction potentials, such as those displayed in Table 6.1, it is necessary to carry out experiments under conditions such that liquid junction potentials

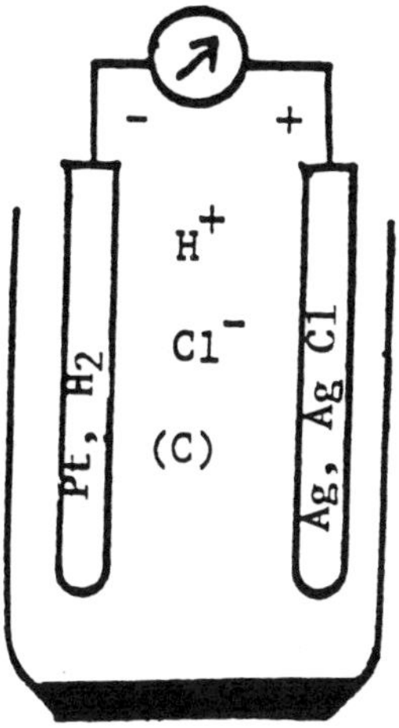

Fig. 6.2. Diagrammatic representation of a cell without liquid junction.

are minimized or even eliminated altogether by using experimental arrangements without liquid junctions.

A widely used cell without liquid junction consists of a hydrogen electrode, a solution of HCl and a silver-silver chloride electrode, consisting of a silver wire or bar coated with solid silver chloride. This cell, illustrated in Figure 6.2, can be represented by

$$\mathrm{Pt} \mid \mathrm{H_2}\ (1\ \mathrm{atm}) \mid \mathrm{HCl}\ (\mathrm{m}) \mid \mathrm{AgCl}\ (\mathrm{s}) \mid \mathrm{Ag} \qquad \text{(cell B)}$$

The reaction for the passage of one faraday of electricity from left to right is

$$1/2\ \mathrm{H_2}\ (\text{gas, 1 atm}) + \mathrm{AgCl}\ (\text{solid}) \rightarrow \mathrm{H^+}\ (\text{aq, m}) + \mathrm{Cl^-}\ (\text{aq, m}) + \mathrm{Ag}\ (\text{solid})$$

Equation 6.1 applied to this reaction is

$$\mathbf{E} = \mathbf{E}^0 - (\mathrm{RT}/\mathbf{F})\ln a_{\mathrm{H}^+}a_{\mathrm{Cl}^-} = \mathbf{E}^0 - (\mathrm{RT}/\mathbf{F})\ln (\gamma_\pm)^2\mathrm{m}^2 \qquad (6.3)$$

Since by definition $\mathbf{V}^0$ for the hydrogen electrode is zero, $\mathbf{E}^0$ for this cell

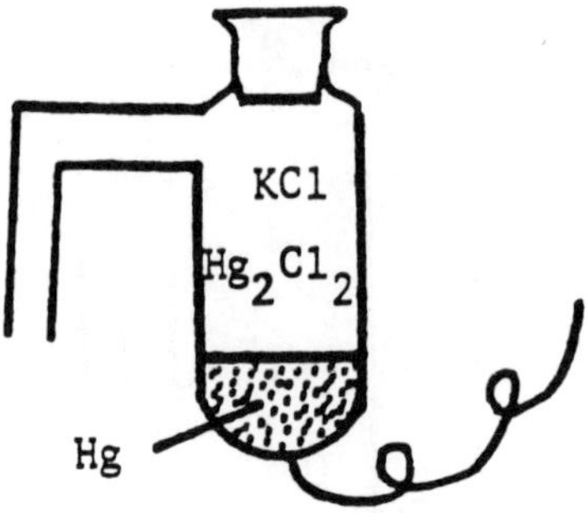

Fig. 6.3. A calomel half-cell.

is the $\mathbf{V}^0$ for the silver-silver chloride electrode. It has a value of +0.222 volts at 25°C.

The calomel electrode or calomel half-cell is frequently used in potentiometric investigations, for example, in the pH meter. This electrode, illustrated in Figure 6.3, consists of a pool of mercury in contact with solid mercurous chloride (calomel) bathed by a solution of potassium chloride. Electrical contact is made through a wire, such as platinum, dipping into the pool of mercury on the one side and a liquid junction between the potassium chloride solution and the system to be measured. The electrode can be represented symbolically as follows:

$$Hg \mid Hg_2Cl_2(s) \mid KCl \mid$$

The electrode potential of the calomel half-cell depends on the concentration of KCl. For 0.10 normal KCl, **V** is −0.3338 volts and for 1.0 normal KCl, it is −0.2801 volts. When the calomel half-cell is used, there is always a liquid junction, and therefore a liquid junction potential, between the KCl solution in the electrode and the electrolyte solution in contact.

LIQUID JUNCTION POTENTIALS

Consider two electrolyte solutions in contact, I on the left and II on the right, as illustrated in Figure 6.4. Let t_i, z_i and a_i be the transference number, the valence and the activity, respectively, of the *i*th ion. When one **F** of positive electricity is passed from left to right, that is, from solution I to solution II, t_i/z_i moles of each ion will move across any

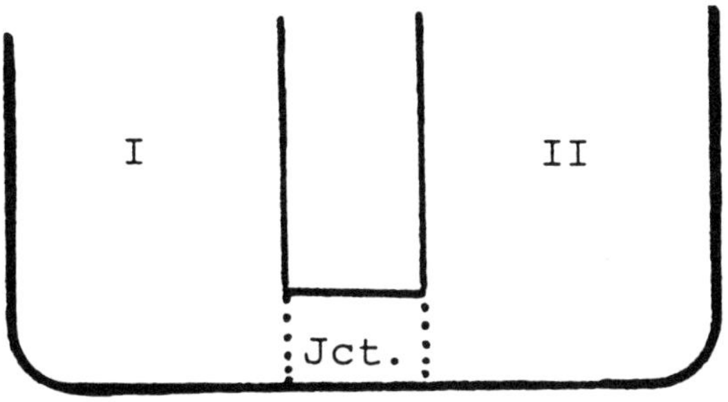

Fig. 6.4. Schematic representation of two electrolyte solutions, I and II, in contact through a liquid junction.

infinitesimally thin layer in the boundary, from left to right if positive and from right to left if negative. The infinitesimal change in Gibbs free energy is given by

$$dG = RT \Sigma (t_i/z_i) \, d\ln a_i$$

Since $\mathbf{E} = -G/\mathbf{F}$ for one faraday,

$$-\mathbf{E}_1 = (RT/\mathbf{F}) \int_{I}^{II} \Sigma (t_i/z_i) \, d\ln a_i \qquad (6.4)$$

The symbol $\mathbf{E}_1$ is the liquid junction potential.

Equation 6.4 can be derived in the following manner. When other sources of potential energy are absent, the potential energy of one mole of an ion, its electrochemical potential, μ_i^*, is the sum of its electrical potential energy, $z_i\mathbf{FE}$, and its chemical potential, μ.

$$\mu_i^* = \mu_i + z_i\mathbf{FE} = \mu_i^0 + RT \ln a_i + z_i\mathbf{FE} \qquad (6.5)$$

When a mole of the *i*th ion is transferred reversibly from a position where its activity is a_i' and the potential is $\mathbf{E}'$ to a position where its activity is a_i'' and the potential is $\mathbf{E}''$,

$$z_i\mathbf{FE}'' + \mu_i'' = z_i\mathbf{FE}' + \mu_i'$$

$$z_i\mathbf{F}(\mathbf{E}'' - \mathbf{E}') = -(\mu_i'' - \mu_i') = -\mathrm{RT}\ \ln a_i'' + \mathrm{RT}\ \ln a_i'$$

$$\mathrm{d}\mathbf{E} = -\frac{\mathrm{d}\mu_i}{z_i\mathbf{F}} = -\frac{\mathrm{RT\ dln}\ a_i}{z_i\mathbf{F}}$$

In this case, z_i faradays are passed. When one faraday is passed, and t_i/z_i moles are transferred,

$$-\mathrm{d}\mathbf{E} = (\mathrm{RT}/\mathbf{F})(t_i/z_i)\ \mathrm{d}\ \ln a_i$$

When more than one ion is involved, the term on the right, being extensive, must be summed over all ions. When the appropriate summation is carried out followed by integration Equation 6.4 results.

The right member of equation 6.4 can be integrated for the general case of two solutions, I and II, each containing several different positive and negative ions only by making several simplifying assumptions. The Henderson integration involves the assumptions that the junction between the two solutions is composed of an infinite number of infinitesimal layers, each containing a solution produced by mixing solutions I and II. It also involves the assumption that μ_i, the mobility of the *i*th ion, is constant and that $\mathrm{dln}\ a_i = \mathrm{dln}\ m_i$, where m_i is the molality of the *i*th ion. This latter is the equivalent of the assumption that the activity coefficient is a constant over the concentration range between solutions I and II, a rather dubious assumption when the two solutions differ substantially. When equation 6.4 is integrated[2] after these assumptions are made, equation 6.6 is obtained.

$$\mathbf{E}_l = [(\mathrm{RT}/\mathbf{F})\Sigma(u_i/z_i)(m_i'' - m_i')/\Sigma(m_i'' - m_i')u_i]\ln(\Sigma m_i'\ u_i/\Sigma m_i''\ u_i) \quad (6.6)$$

Beyond the problem of the assumptions made in the derivation of equation 6.6 from equation 6.4, there is an enormous amount of information necessary to evaluate the liquid junction potentials involving junctions with solutions containing many ions, as is common in biological systems. Equation 6.6 can be simplified for certain special cases [1]:

1. If all ions are univalent, z_i is $+1$ for positive ions and -1 for

negative ions. When the following definitions are made and then substituted into equation 6.6,

$$U_1 \equiv \Sigma m_i' u_i' \text{ and } U_2 \equiv \Sigma m_i'' u_i'' \text{ for positive ions}$$
$$V_1 \equiv \Sigma m_i' u_i' \text{ and } V_2 = \Sigma m_i'' u_i'' \text{ for negative ions}$$

$$\mathbf{E}_1 = (RT/\mathbf{F})[(U_1 - V_1) - (U_2 - V_2)]/[(U_1 + V_1) - (U_2 + V_2)] \ln (U_1 + V_1)/U_2 + V_2) \quad (6.7)$$

2. If solutions I and II are a single uni-univalent salt at concentrations m′ and m″, respectively, then equation 6.6 becomes

$$\mathbf{E}_1 = (RT/\mathbf{F})[(u^+ - u^-)(m'' - m')/(u^+ + u^-)(m'' - m')]\ln m'(u^+ + u^-)/m''(u^+ + u^-)$$

$$\mathbf{E}_1 = (RT/\mathbf{F})\,[(u^+ - u^-)/(u^+ + u^-)]\ln m'/m'' \quad (6.8)$$

Since $t^+ = u^+/(u^+ + u^-)$ and $t^- = u^-/(u^+ + u^-)$ and $t^- = 1 - t^+$, equation 6.8 can be transformed into

$$\mathbf{E}_1 = (2t^+ - 1)(RT/\mathbf{F})\ln m'/m'' = (t^+ - t^-)\,(RT/\mathbf{F})\ln m'/m'' \quad (6.9)$$

3. When solutions I and II are two uni-univalent salts at concentration m with a common ion, equation 6.6 becomes

$$\mathbf{E}_1 = (RT/\mathbf{F})[m(u^{+''} - u^{+'} - u^{-''} + u^{-'})/m(u^{+''} + u^{-''} - u^{+'} - u^{-'})] \ln(u^{+'} + u^{-'})/(u^{+''} + u^{-''})$$

When $u^{-''} = u^{-'}$, that is, when the negative ion is the common ion,

$$\mathbf{E}_1 = (RT/\mathbf{F}) \ln (u^{+'} + u^{-'})/(u^{+''} + u^{-''}) = (RT/\mathbf{F}) \ln \lambda'/\lambda'' \quad (6.10)$$

In this equation, λ is the equivalent conductance at the concentration, m.

Equation 6.4 can be integrated more simply for two special cases. The first is that of a single symmetrical electrolyte at concentration m′ on the left and m″ on the right. The necessity to preserve electrical neutrality requires that, in every region of the boundary, the concentrations of the positive and negative charges be the same and be equal to zm, where z is the absolute magnitude of the valence of both ions. Since $t^+ = m^+u^+/(m^+u^+ + m^-u^-) = u^+/(u^+ + u^-)$ and $t^- = u^-/(u^+ + u^-)$, t^+ and t^- will be constants provided only that u^+ and u^- vary the same way with concentration. This is not a drastic assumption. While there is no way to measure the activity coefficient, γ_i, of a single ion, useful results are obtained in many situations by assuming that γ_i equals $\gamma_\pm$, the mean activity coefficient (see Eq. 4.20). By this assumption $a_+ = \gamma_\pm m = a_- = a_i$. Thus, when the limits of integration are inverted, equation 6.4 becomes

$$\mathbf{E}_1 = (RT(t^+ - t^-)/z\mathbf{F})\int_{II}^{I} d\ln a_i = (RT(t^+ - t^-)/z\mathbf{F})\ln (m'/m'')(\gamma'_\pm/\gamma''_\pm) \quad (6.11)$$

When $\gamma'_\pm/\gamma''_\pm$ and $|z|$ are each 1, this is the same as equation 6.9.

The second special case is when the two electrolytes are separated by a membrane permeable to one ion only. In that case, t is 1 for the permeable ion and 0 for the other. For either ion, when the sign of z_i is retained,

$$-\mathbf{E}_1 = (RT/z_i\mathbf{F})\int_{I}^{II} d\ln a_i, \text{ or}$$

$$\mathbf{E}_1 = (RT/z_i\,\mathbf{F})\ln \gamma'_\pm m'/\gamma''_\pm m'' \quad (6.12)$$

This equation is useful for the interpretation of the potentials across some biological membranes. A hypothetical example fitting the equation would be a membrane with small charged pores that allowed only ions of opposite sign to penetrate.

CONCENTRATION CELLS WITHOUT TRANSFERENCE

Consider two cells like cell B that are identical except that the concentration of HCl in the cell on the left is m′ and on the right is m″.

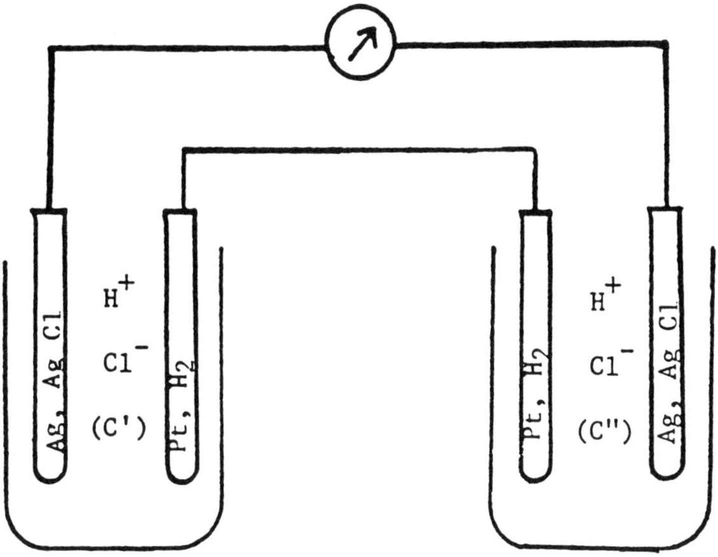

Fig. 6.5. Schematic representation of a concentration cell without transference.

In both, 1 mole of ionized HCl is produced when one faraday is passed from left to right. On the left, by equation 6.1, $\mathbf{E}' = \mathbf{E}^{\circ} - (RT/\mathbf{F}) \ln a'_{H+} a'_{Cl-}$. On the right, $\mathbf{E}'' = \mathbf{E}^{\circ} - (RT/\mathbf{F}) \ln a''_{H+} a''_{Cl-}$.

If now, as shown in Figure 6.5, the cell on the left is turned around so that the hydrogen electrode is on its right and if the leads from the two hydrogen electrodes are connected so that they are at the same potential, **E** for the combined cells will be given by

$$\mathbf{E} = \mathbf{E}'' - \mathbf{E}' = -(RT/\mathbf{F}) \ln a''_{H+} a''_{Cl-} + (RT/\mathbf{F}) \ln a'_{H+} a'_{Cl-}$$

$$\mathbf{E} = RT/\mathbf{F} \ln (\gamma'_{\pm})^2 (m')^2/(\gamma''_{\pm})^2 (m'')^2 \qquad (6.13)$$

When 1 faraday is passed, 1 mole of ionized HCl will be removed from the solution on the left, and one mole will be produced in the solution on the right. Thus one mole of ionized HCl will be transferred from the left solution to the right solution, not by direct migration across a liquid junction, but by the reactions at the electrodes.

The cell just described could be written

$$Ag \mid AgCl \mid HCl\ (m') \mid H_2 \mid HCl\ (m'') \mid AgCl \mid Ag$$

If, instead of HCl and H_2, Cu Cl_2 and Cu were substituted and the Cu were the barrier between the solutions, the cell would be

$$Ag \mid AgCl \mid CuCl_2\ (m') \mid Cu \mid CuCl_2\ (m'') \mid AgCl \mid Ag$$

When 1 **F** is passed, 1/2 mole of Cu^{2+} and 1 mole of Cl^- will be removed from the left and will be produced on the right. Equal amounts of copper will be deposited on the left surface and removed from the right surface of the copper barrier separating the two solutions. Application of equation 6.1 results in

$$\mathbf{E} = RT/\mathbf{F} \ln a'^{1/2}_{Cu2+}\ a'^{1}_{Cl-} \ /\ a''^{1/2}_{Cu2+}\ a''_{Cl-}$$

$$a^{1/2}_{Cu2+}\ a_{Cl-} = \gamma_+^{1/2}\ \gamma_-\ m_+^{1/2}\ m_- = \gamma_+^{1/2}\ \gamma_-\ m^{1/2}\ 2\ m$$

By equation 4.20, $\gamma_\pm = (\gamma_+^{1/2}\ \gamma_-)^{2/3}$. Thus,

$$\mathbf{E} = (RT/\mathbf{F}) \ln (\gamma'_\pm\ m')^{3/2}/(\gamma''_\pm\ m'')^{3/2} = (3RT/2\mathbf{F}) \ln \gamma'_\pm\ m'/\gamma''_\pm\ m'' \quad (6.14)$$

In a sense, Cu^{2+} ions would be passing through a copper "membrane."

CONCENTRATION CELLS WITH LIQUID JUNCTION

If now the hydrogen electrodes are removed from the two cells described in Figure 6.5 and, as shown in Figure 6.6, a liquid junction is established between the two solutions, the electrode reactions will involve the chloride ion only. When 1 faraday is passed from left to right, 1 equivalent of Cl^- ion will be removed at the electrode on the left and 1 will be added at the electrode on the right. In addition, t^+ equivalents of H^+ ion will move from the left solution to the right solution and t^- equivalents of Cl^- ion will move from the right solution to the left. The net loss of Cl^- ion from the left solution will be $(1 - t^-) = t^+$ equivalents, and the net gain of Cl^- in the right solution will be $(1 - t^-) = t^+$ equivalents. Thus t^+ moles of ionized HCl will be transferred from the left solution to the right solution. For the whole operation,

$$\mathbf{E} = (RT/\mathbf{F}) \ln a'_{Cl-}/a''_{Cl-} + \mathbf{E}_l.$$

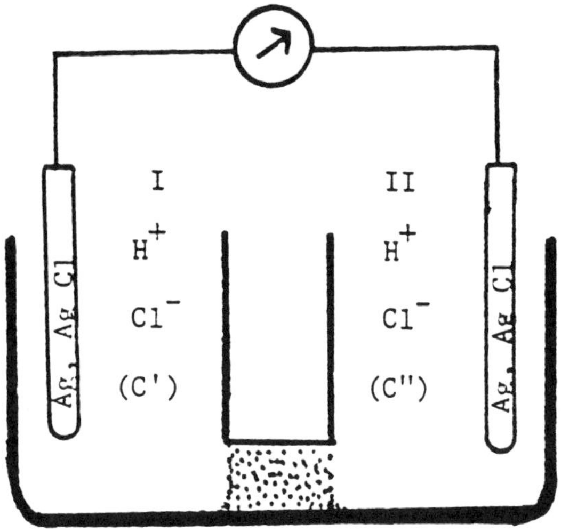

Fig. 6.6. Schematic representation of a concentration cell with a liquid junction.

By substituting equation 6.11, since $|z|$ is unity and $t^+ - t^- = 2t^+ - 1$,

$$\mathbf{E} = (\mathrm{RT/F}) \ln (\gamma_\pm)' \mathrm{m}'/\gamma''_\pm \mathrm{m}'' + (2t^+ - 1)(\mathrm{RT/F}) \ln (\gamma_\pm)' \mathrm{m}'/(\gamma_\pm)'' \mathrm{m}''$$

$$\mathbf{E} = (2t^+ \mathrm{RT/F}) \ln (\gamma_\pm)' \mathrm{m}'/(\gamma_\pm)''\mathrm{m}'' \quad (6.15)$$

The potential from right to left across the cell will also be **E**. Thus, if m′ is greater than m″, the dilute solution will be positive to the concentrated one.

If hydrogen electrodes are substituted for the silver-silver chloride electrodes in Figure 6.6, when 1 faraday is passed from left to right, one equivalent of H^+ ion will be produced at the electrode in the solution at the left, where the HCl concentration is m′, and one equivalent will be removed at the electrode from the solution on the right. At the liquid junction, t^+ equivalents of H^+ ion will move from the left solution to the right and t^- equivalents of Cl^- will cross the boundary in the opposite direction. The net result will be an increase of $(1 - t^+) = t^-$ equivalents of H^+ ion and t^- equivalents of Cl^- ion in the left solution and a decrease of t^- equivalents of both ions in the right solution. Since

t^- moles of HCl are transferred from right to left in this case, instead of from left to right as in the preceding paragraph, the liquid junction potential must be subtracted from the net electrode potential. For the whole cell,

$$\mathbf{E} = + (\mathrm{RT}/\mathbf{F}) \ln (\gamma_\pm)'\mathrm{m}'/(\gamma_\pm)''\mathrm{m}'' - (2t^+ - 1)(\mathrm{RT}/\mathbf{F}) \ln (\gamma_\pm)'\mathrm{m}'/(\gamma_\pm)''\mathrm{m}''$$

Since $t^+ - 1 = -t^-$,

$$\mathbf{E} = + 2t^- (\mathrm{RT}/\mathbf{F}) \ln (\gamma_\pm)'\mathrm{m}'/(\gamma_\pm)''\mathrm{m}'' \qquad (6.16)$$

It should be noted from this and the preceding deduction that the transference number to be used is the one to which the electrodes are *not* reversible.

This discussion of concentration cells with liquid junction potentials presents a logical dilemma. Liquid junction potentials are real, but they are not reversible. Thus deductions about liquid junction potentials in terms of thermodynamic consideration cannot be rigorous. Yet Hitchcock [2] reported measurements of **E** for the concentration cell 0.10 normal HCl | 0.01 normal HCl, when H_2 electrodes are used to be 19.3 millivolts and when Ag, AgCl electrodes are used to be 92.5 mV. Since $\mathbf{E}_1$ is subtracted from the net electrode potential in the first case and added in the second, the simultaneous equation $\mathbf{E} - \mathbf{E}_1 = 19.3$ and $\mathbf{E} + \mathbf{E}_1 = 92.5$ can be solved to yield $\mathbf{E} = 55.9$ mV and $\mathbf{E}_1 = 36.6$. mV. Examination of equation 6.9 shows that $\mathbf{E}_1/\mathbf{E}$ should be $(2t^+ - 1)$. The value of t^+ for the H^+ ion in a solution of HCl is about 0.83. Thus $\mathbf{E}_1/\mathbf{E}$ should be 0.66. The experimentally determined ratio is 0.65. Despite the lack of rigor, the treatment seems to be satisfactory.

The only thing that happens at a junction between solutions of the same electrolyte at different concentrations is diffusion from the concentrated solution to the dilute solution. The mobility and, therefore, the diffusion rate, of the hydrogen ion is much higher than that of the chloride ion. Thus it tends to move toward the dilute solution more rapidly than the chloride ion. Electrical neutrality must be preserved. A potential negative on the concentrated side is thereby generated. Hence the liquid junction potential is in reality the diffusion potential. Diffusion of electrolytes and diffusion potentials were discussed at greater length in Chapter 3.

If now calomel electrodes are substituted for the hydrogen electrodes in vessels I and II, Figure 6.6., the total **E** for the cell will be $\mathbf{E}_{cal}$ + $\mathbf{E}_{l1}$ + $\mathbf{E}_{l}$ − $\mathbf{E}_{l2}$ − $\mathbf{E}_{cal}$, where $\mathbf{E}_{l1}$ and $\mathbf{E}_{l2}$ are the liquid junction potentials between the KCl bridges of the calomel electrodes and solutions I and II, respectively. If it is assumed the $\mathbf{E}_{l1}$ and $\mathbf{E}_{l2}$ are approximately equal, by equation 6.11,

$$\mathbf{E} = \mathbf{E}_l = (2t^+ - 1)\,RT/z\,\mathbf{F}\ \ln \gamma'_{\pm}m'/\gamma''_{\pm}m'' \qquad (6.17)$$

The liquid junction or diffusion potential is measured directly in this system. It should be remembered that diffusion potentials are not equilibrium potentials. As equilibrium is approached, m′ and m″ approach equality, and the potential approaches zero.

It is clear from this discussion that the measured potential in a cell containing a liquid junction depends critically upon the electrode used. Only when irreversible or inert electrodes are used does one measure directly the true liquid junction or diffusion potential.

An interesting possibility for biological transport is afforded by liquid junction potentials even in the absence of membranes. As a hypothetical case, suppose that somewhere within a cell a high molecular weight slow diffusing or almost immobile material is synthesized and suppose that this material dissociates hydrogen ions. Suppose further that somewhere else in the cell with no membrane intervening, there is a hydrogen ion sink. In the absence of convection, a hydrogen ion gradient will be established between these two points. This will generate a diffusion potential, negative in the direction of the hydrogen ion source. This potential will act on all other ions in the cell, facilitating the movement of positive ions or positively charged macromolecules toward the hydrogen ion source and negative ions or negatively charged particles toward the hydrogen ion sink. This resembles somewhat the happenings leading toward Donnan equilibrium, but without a membrane. In a sense, an ion exchange process will be in operation. The events at the source and at the sink would be the result of metabolic activity and would be active, but the rest of the events would be passive, subject to the rules of diffusion and of migration of charged particles in an electric field.

SALT BRIDGES

Salt bridges are widely used in electrophysiology and in common laboratory pH meters. Wherever there is a salt bridge there is of

TABLE 6.2. Calculated Values of E_1 at 25°C

I	II	E_1 (volts)
0.01 M HCl	0.10 M HCl	−0.038
0.01 M KCl	0.10 M KCl	+0.0011
0.10 M KCl	4 M KCl	+0.0018
0.01 M HCl	4 M KCl	+0.0030
0.10 M HCl	4 M KCl	+0.0047
0.10 M HCl	0.15 M NaCl	+0.027

necessity a liquid junction potential. Physical scientists avoid liquid junctions whenever possible when they carry out experiments that demand the utmost in precision. But in biology salt bridges are frequently used as a matter of necessity, as is pointed out elsewhere in this chapter. Equilibrium potentials, when they exist, cannot be measured with reversible electrodes; the salt bridge is a convenient nonreversible alternative means of making electrical contact.

It is common knowledge that liquid junction potentials are minimized when KCl solutions are the salt bridge. But what might be forgotten is that liquid junction potentials between KCl bridges and electrolyte solutions are not zero—they can amount to several millivolts.

The reason why KCl solutions used as salt bridges give reasonably low values of liquid junction potential can be understood in terms of equation 6.6. When the junction between two concentrations of the same uni-univalent electrolyte is being considered, equation 6.8, a modified form of equation 6.6 appropriate for this situation, can be used. Ionic mobilities, u, are related to ionic conductance by the equation, $\lambda = \mathbf{F}u$. Since ionic mobilities are involved to the same degree in the numerator and denominator of equation 6.8, ionic conductance can be substituted. Limiting values of the ionic conductance, as concentration approaches 0, for the H^+, the K^+, the Na^+, and the Cl^- ions are 349.8, 73.5, 50.1 and 76.3 reciprocal ohms, respectively. When these values are used for concentrated solutions, the results can be approximations only, but they can be of use in illustrating principles. Table 6.2 gives the results of calculations of $\mathbf{E}_1$ for various liquid junctions between solutions I and II. It should be noted that the calculated value of $\mathbf{E}_1$ for the junction between 0.01 M HCl and 0.10 M HCl is −0.038 volts, which happens to agree well with the experimentally determined value mentioned elsewhere. In contrast, when the boundary is between 0.01 M KCl and 0.10 M KCl, the calculated value of the liquid junction potential is only 0.0011 volts. This is in accord with the view that liquid junction potentials are minimized by using KCl solutions. The reason for the difference is that the ionic conductance of H^+ is very much larger than that of Cl^-, while that of K^+ is more nearly equal that of Cl^-. Hydrogen ions tend to diffuse much faster than chloride ions; a large potential develops. In the case of the K^+ ion, its conductance and therefore its mobility is just slightly less than of Cl^- ion. This leads to a low diffusion potential in the opposite direction. Table 6.2 shows further that even when the liquid junction is between 0.10 M KCl and 4 M KCl the potential is still only 0.0018 volts. The calculated results for boundaries between 0.01 M HCl or 0.10 M HCl and 4 M KCl are 0.0030 and 0.0047 volts, respectively. The variation shown in Table 6.2 between approximately 2 and approximately 5 mV volts for the liquid junction potential between 4 M KCl, approximately saturated KCl, and the other electrolytes shown illustrates the importance of liquid junction potentials. When physiological saline, 0.15 M NaCl, is part of the liquid junction, the potentials are much higher. If a potential of only a few mV is recorded between two solutions or two points on a tissue or between the inside and outside of the cell, the result could be meaningless, amounting to nothing more than different liquid junction potentials at the two points of contact. Fortunately, the potentials observed in biological systems and across artificial membranes are frequently at least an order of magnitude greater than these liquid junction potentials and therefore do have significance. However, they must of necessity have limited accuracy when salt bridges are used to make electrical contact.

MEMBRANE POTENTIALS

If solutions I and II are separated by a membrane equally permeable to all ions, but, for the sake of simplicity, impermeable to solvent, the situation will be the same as that described for the two solutions connected by a liquid junction except that equilibration will take place more slowly. The recorded potential difference between the two sides of the membrane will depend upon what kind of electrode is used. If nonreversible electrodes like the calomel half-cell are used, the diffusion potential will be measured directly provided only that the liquid junction potentials between the KCl bridges of the two calomel electrodes and their respective solutions are nearly equal. Equation 6.11 will apply if a single electrolyte such as HCl is found on both sides of the membrane. If more complex solutions are in vessels I and II, equation 6.6 will apply approximately.

If the membrane is permeable to one kind of ion only, equation 6.12 will apply. The liquid junction potential measured when calomel or other nonreversible electrodes are used will depend solely upon the ratio of the activities of that ion in vessels I and II. A simple way to visualize a membrane permeable to ions of positive charge only is to think about very small pores with surface charges. If the surface charge is negative and if the pore is small enough, then negative ions in the solution will be repelled so strongly that they cannot penetrate the pore. Positive ions can enter the pore and thus pass through it to some extent. If the pores are very small and have a positive charge, then positive ions will be excluded, but the membrane could be slightly permeable to negative ions. Weak binding centers in the pores for one kind of ion only could also make a membrane permeable to that ion only.

As ion exchange membrane conditioned with a particular ion (i.e., equilibrated with that ion) can be used as a reversible electrode for that ion, provided that no interfering ions of the same charge are present in the system. When membrane adsorbs a particular ion, the potential between the membrane and the adjacent solution is sensitive to the concentration of that ion. Scatchard et al. [3,4] have used such membranes for the purpose of measuring ion binding by proteins, and Shalaby et al. [5] measured ion binding by tobacco mosaic virus and its protein in a similar manner. As was previously pointed out, a metal behaves like a membrane permeable to its ion. Equation 6.12 will apply when nonreversible electrodes like the calomel half-cell are used to make contact with the solutions on both sides of the metal barrier, subject to the provision mentioned above.

The glass electrode was once considered a membrane with selective permeability for hydrogen ions. This simple interpretation is not rigorously acceptable. A more nearly satisfactory conceptual basis for understanding its operation is that it is something akin to an ion exchange membrane.

Beutner [6] has reported a variety of experiments in which electrolyte solutions, each in contact with a calomel electrode, are separated by water immiscible liquids, organic compounds that contain OH, NO_2, CO, or COH groupings. Potential differences of as much as several hundred mV can be observed in such systems. When 0.1 M KCl is separated from 0.0008 M KCl by salicylic aldehyde containing considerable amounts of salicylic acid, the observed potential was 0.10 volt. This is not drastically different from the value that would be calculated by equation 6.12 on the assumption that the organic liquid is acting as a membrane permeable to one ion only.

Beutner also reports results obtained when KCl and NaCl solutions are separated by a layer of phenol or cresol. The KCl is negative with respect to the NaCl. If a 0.1 M NaCl solution is separated by cresol from a 0.1 M sodium oleate solution, the NaCl solution is about 1/10 volt negative to the other. When 0.1 M NaCl solution is separated by cresol from 0.1 M aniline hydrochloride, the latter is negative by 0.08 volts. In both of these cases, the organic ions are more soluble in the organic liquid than are the inorganic ions.

A qualitatively plausible explanation of these phase boundary potentials, influenced strongly by the views of Höber [7], is that the organic liquids, phenol, cresol, and salicylic aldehyde contaminated by salicylic acid, are cation exchangers. Thus, the result obtained when the organic liquid separated two KCl solutions of different concentrations can be explained on the theory that the organic liquid is permeable to K^+ ion but not to Cl^- ion. Equation 6.12 should apply. The potential between the two solutions is really the sum of two interfacial potentials, one between the dilute solution and the organic liquid and the other between the concentrated solution and the organic liquid. These two potentials depend upon the potassium ion concentration in the organic liquid. This is assumed to be low and the same at both interfaces; the ion exchange capacity of the small amount of contaminating salicylic acid is saturated at both interfaces. The result obtained when sodium chloride and potassium chloride at the same concentration are separated by cresol requires a further assumption. The sodium ion is generally regarded as being significantly more hydrated than the potassium ion. This causes the penetration of Na^+ into the cresol to be less than that of K^+. Therefore the interfacial potential on the Na^+ side is of greater negative magnitude than that on the K^+ side. However, since the greater negative magnitude is in the opposite direction from the smaller, it results in a potential that is positive on the Na^+ side.

The potentials that exist across membranes permeable to a single ion only are equilibrium potentials. They can be thought of as the potential just necessary to prevent the more rapidly diffusing ion from being separated to any great distance from the more slowly diffusing ion of opposite charge. Since the potential at the membrane is an equilibrium potential, if electrodes reversible to the penetrable ion are inserted in the solution, the potential of the system will be zero because the whole system is at equilibrium and incapable of doing work. It is therefore necessary to use nonreversible or inert electrodes to measure these membrane potentials.

DONNAN POTENTIALS

As was discussed in Chapter 1, when a solution, I, containing a low molecular weight electrolyte and a charged biopolymer, such as the salt of a protein, is separated from a solution, II, of the low molecular weight electrolyte by a membrane permeable to low molecular weight ions and solvent but impermeable to the charged biopolymer, Donnan equilibrium will be established. The condition for equilibrium is that vessel I be at a higher pressure than vessel II and that the concentration of ions of charge with the same sign as that of the biopolymer be at a higher magnitude inside vessel II than in vessel I and the opposite for ions of the opposite charge. It was shown further that equation 1.6, the Donnan equilibrium equation, can be written in the form $m'_+ \, m'_- = m''_+ \, m''_-$ where single primes refer to solution I, double primes to II, and m means molality. This can be rewritten in the form $m'_+/m''_+ = m''_-/m'_-$. In a solution involving many permeable ions, positive and negative, the same equation must hold for each. If electrodes reversible to one of the ions are placed in each of the solutions, it can readily be deduced from thermodynamics that no potential difference can be observed. The reason is simply that equilibrium exists at all points in the system, at both electrodes and at the membrane. It is thermodynamically impossible for a system at equilibrium to do work. Since an external potential difference at the electrodes would allow a current to flow and do work, this cannot happen. At the turn of the century, Jacques Loeb [8] carried out many experiments on Donnan equilibrium systems involving collodion membranes impermeable to proteins but permeable to all other constituents. He demonstrated experimentally that no potential can be measured in the setup just described, in agreement with thermodynamic deduction. Yet, if an ion-specific reversible electrode and a calomel reference electrode are placed in each of the two solutions, differences in the two potentials should be and are observed, because differences in the concentrations, and therefore in the activities, of the electrolyte in equilibrium with the reversible electrode exist between solution I and II. Since no external potential is observed when the reversible electrodes in the two solutions are connected, and yet the electrode potentials must be different, there must be at equilibrium a potential across the membrane equal to but opposite to the difference between the reversible electrode potentials. This is the Donnan potential.

For the sake of specificity, suppose that the impermeable ion has a negative charge, that the chloride ion is one of the permeable ions, and that Ag | AgCl electrodes, reversible to the Cl^- ion are used. In this case m''_-/m'_- will be greater than 1. The difference between the electrode potentials should be $(RT/\mathbf{F})\ln a'_-/a''_-$, that is, the electrode potential in II should be more negative than in I. Since the measured value is 0, the Donnan potential at the membrane must be of the same magnitude but opposite in direction. That is, the side of the membrane adjacent to solution I containing the impermeable negative ion must be negative and that facing solution II must be positive.

$$\mathbf{V}_d = (RT/\mathbf{F}) \ln a''_-/a'_- \simeq (RT/\mathbf{F}) \ln m''_-/m'_- = (RT/\mathbf{F}) \ln m'_+/m''_+ \qquad (6.18)$$

$\mathbf{V}_d$ is the Donnan potential.

The existence of this potential across the membrane accounts for the concentration of positive ions being greater at equilibrium in solution I than in solution II and the reverse for negative ions. Loeb showed that if nonreversible electrodes, such as calomel electrodes, are connected to both solutions I and II, then a potential difference of the right sign and the right magnitude can be observed. If in biological systems the conditions for the establishment of a Donnan equilibrium exist, thermodynamics being what it is, there must be Donnan potentials across the membranes.

RESTING POTENTIALS IN BIOMEMBRANES

The next issue that needs to be considered is the source of biological membrane potentials. The inside of the cell contains substantial amounts of macromolecular materials, especially proteins and nucleic acids, which are negatively charged at physiological pH values. The cell membrane can be thought to be essentially impermeable to these negatively charged macromolecules. The situation is thus one required for a Donnan potential, provided that the membrane is permeable to at least one electrolyte. It has been known for a long time that biological membranes are highly permeable to K^+ ions and Cl^- ions but not to

Na^+ ions. In the discussion of Donnan potentials in the previous section, it was shown in equation 6.18 that

$$V_d = -(RT/F)\ln a'_-/a''_- = -(RT/F)\ln a''_+/a'_+$$

The second equality of that equation stems from the equality of a'_-/a''_- and a''_+/a'_+, discussed in Chapter 1, equation 1.6. A well-investigated cell is the squid giant axon. The potassium ion concentration inside the cell is about 0.4 M and outside about 0.02 M. When these values are substituted into equation 6.18, a Donnan potential of 0.077 volts is obtained. This is of the order of magnitude of the measured resting potential of the giant axon, about 0.060 volts, but it is, nevertheless, significantly higher. In addition to the fact that the simple Donnan theory doesn't give quite the right answer, there are other difficulties. First of all, the ratio of the potassium concentration inside to that outside should be equal to the ratio of the chloride concentration outside to that inside. The chloride ion concentration on the outside is about 0.56 M and that on the inside ranges between 0.04 and values several times higher. Thus the ratio is not the same as that for potassium even for the smallest of the above-mentioned inside concentrations. Furthermore, permeability studies have shown that while Na^+ ions penetrate the membrane much more slowly than K^+ ions, its permeability is not zero. Similar studies have established that the permeability of the chloride ion, while much greater than that of the sodium ion, is definitely less than that of the potassium ion.

A more sophisticated version of the Donnan theory is required by the facts recorded above. In a true Donnan equilibrium,

$$m''_{Na^+}/m'_{Na^+} = m''_{K^+}/m'_{K^+} = m'_{Cl^-}/m''_{Cl^-} =$$

$$m'_{X^-}/m''_{X^-} = (\Sigma m''_+ + \Sigma m'_-)/(\Sigma m'_+ + \Sigma m''_-).$$

If the membrane is only partially permeable to a given ion, the situation is as though only a fraction of the area is fully permeable. Thus the effective concentration of each ion is proportional to the product of its permeability, P_i, multiplied by its molality, m_i. The Donnan theory can be modified by substitution $(\Sigma P_i\, m''_+ + \Sigma P_i\, m'_-)/(\Sigma P_i\, m'_+ + \Sigma P_i\, m''_-)$ for m''_+/m'_+ in equation 6.18.

$$V_d = -(RT/F)\ln(\Sigma P_i\, m''_+ + \Sigma P_i\, m'_-)/(\Sigma P_i\, m'_+ + \Sigma P_i\, m''_-) \quad (6.19)$$

This equation was derived by D.E. Goldman in 1943 and is thus frequently called the Goldman equation. The permeability of Na^+ is taken as 0.04 relative to that of K^+ and that of the Cl^- ion is 0.45 relative to that of K^+; when the concentration of Cl^- inside is taken to be 0.05 M, equation 6.19 yields a value of +0.061 volts, a value in satisfactory agreement with the experimentally measured value. The membrane is positive on the outside relative to the inside.

The Goldman equation provides a reasonably satisfactory basis for estimating the resting potential in terms of a purely passive process. It is also reasonably successful in explaining the ratio of positive ion concentrations inside to outside to that of chloride ion concentrations outside to inside, when relative permeabilities are taken into account. However, it does not explain why the potassium ion concentration inside is about double that which would be calculated from the resting potential alone and the sodium ion concentration inside is vastly lower than that outside. An enzyme isolated by J.C. Skou called sodium-potassium ATPase is a transmembrane protein. It hydrolyzes ATP to ADP only in the presence of both Na^+ and K^+ and pumps Na^+ out of the cell and K^+ into it. The energy is supplied by the hydrolysis of ATP, the thermodynamics of which is discussed in Chapter 7. The postulated action of this transmembrane enzyme is that inside the cell the transmembrane enzyme in one conformational state binds three sodium ions coupled with the hydrolysis of one ATP and the phosphorylization of the enzyme. Magnesium ion is required to catalyze this reaction. Outside the cell the phosphorylated enzyme-sodium ion complex dissociates three Na^+ ions. The phosphorylated enzyme undergoes a conformational change during this dissociation and then binds two K^+ ions. The potassium-phosphorylated enzyme then becomes dephosphorylated on the inside and dissociates its two K^+ ions and reverts to the original conformation on the inside, in preparation for beginning a new cycle. While many authors have thought up imaginative cartoons to illustrate how this enzyme might accomplish this action, the exact mechanism remains unknown. Suffice it to say that the enzyme itself does not move back and forth within the membrane. This sodium-potassium pump is a metabolic process, an example of active transport. Thus the resting potential of the membrane is partially explained by a modified form of the Donnan potential equation, to that extent passive, and partially by the action of the sodium-potassium pump, to that extent active.

Ling [9], whose painstaking work is rarely cited, has vigorously challenged the concept that the sodium-potassium pump is responsible for the unequal ionic concentrations inside and outside the cell membrane. Among other reasons he puts forth is the fact that the energy required to operate the pump continuously vastly exceeds the energy available to the cell. He interprets the accumulation of K^+ and exclusion of Na^+ ions in terms of the association-induction (A-I) hypothesis. According to this hypothesis, K^+ ions are selectively adsorbed by proteins inside the cell. The role of ATP is to enhance selective K^+ adsorption. Na^+ and other ions are excluded because most of the water within the cell is "bound" water, incapable of dissolving electrolytes to the same degree as "free" water. (See Note 3 of Chapter I.) Critical evaluation of the relative merits of these competing theories is beyond the range of expertise of the author, but it is his view that the A-I hypothesis is thermodynamically feasible.

Ling proposed a related theory for the resting potential, the surface adsorption (S-A) theory. The outer surface of the cell membrane, according to the S-A theory, also contains groups that selectively adsorb K^+ ions strongly and Na^+ ions less strongly. This converts the external cell surface into something akin to a cation specific electrode responsive to external cation concentrations, expecially K^+ and, to a lesser degree, Na^+. Ling interprets much experimental evidence to be consistent with the S-A theory and not consistent with equation 6.19 and other formulations of the membrane theory. In calling attention to Ling's challenge the author desires only to alert the reader to the existence of a seriously supported alternative interpretation.

REFERENCES

1. MacInnes, D.A. The Principles of Electrochemistry. New York: Reinhold Publishing Corp., 1939.
2. Hitchcock, D.I. In Physical Chemistry of Cells and Tissues, R. Höber, (ed.). Philadelphia: Blakiston, pp. 62–63.
3. Scatchard, G., Coleman, J.E., and Shen, A.L. J. Am. Chem. Soc. *79*:12–20, 1957.
4. Scatchard, G., Wu, Y.V., and Shen, A.L. J. Am. Chem. Soc. *81*:6104–6109, 1959.
5. Shalaby, R.A., Banerjee, K., and Lauffer, M.A. Biochemistry *7*:955–960, 1968.
6. Beutner, R. Physical Chemistry of Living Tissues and Life Processes. Baltimore: Williams and Wilkins, 1933.
7. Höber, R. Physical Chemistry of Cells and Tissues. Philadelphia: Blakiston, 1945, pp. 285.

8. Loeb, J. Proteins Colloidal Behavior. New York: McGraw-Hill, 1922.
9. Ling, G.N., *In Search of the Physical Basis of Life,* Plenum Press, NY 1984.

NOTES

1. At constant temperature, for the chemical reaction,

$$aA + bB \rightleftarrows cC + dD$$

$$\Delta G = -RT \ln K_a + RT \ln (a_C^c\, a_D^d / a_A^a\, a_B^b).$$

When no nonmechanical work is done, the condition for reversibility or equilibrium is $\Delta G = 0$. When the only nonmechanical work is the electrical work done by the transfer of n equivalents of electrons, for a reversible process, $\Delta G = -W_{el}$ or

$$\Delta G = -n\mathbf{FE}$$

$$\mathbf{E} = (RT/n\mathbf{F}) \ln K_a - (RT/n\mathbf{F}) \ln (a_C^c\, a_D^d / a_A^a\, a_B^b)$$

$$\mathbf{E}^o \equiv (RT/n\,\mathbf{F}) \ln K_a$$

$$\mathbf{E} = \mathbf{E}^o - (RT/n\mathbf{F}) \ln (a_C^c\, a_D^d / a_A^a\, a_B^b)$$

2. Each infinitesimal layer in the boundary contains n kinds of ion, 1, 2, 3 . . . n. If m_i' and m_i'' and the molalities of the *i*th ion in solutions I and II, respectively, and x is the fraction of solution II and $(1 - x)$ the fraction of solution I in a given thin layer in the boundary, the concentration, m_i, of the *i*th ion in that layer is given by $m_i = m_i' + (m_i'' - m_i')\,x$, where x varies from 0 in solution I to 1 in solution II. Similarly for any ion, j,

$$m_j = m_j' + (m_j'' - m_j')\,x.$$

Since t_i is the fraction of the current carried by the *i*th ion,

$$t_i = m_i\, u_i / \Sigma m_j\, u_j = m_i\, u_i / (\Sigma m_j'\, u_j + x\Sigma (m_j'' - m_j')\, u_j)$$

$$t_i = m_i\, u_i/(B_1 + B_2\, x)$$

where $B_1 \equiv \Sigma m_j'\, u_j$ and $B_2 \equiv \Sigma(m_j'' - m_j')\, u_j$.

This simplification involves the assumption that all values of u_j, including u_i, are constants independent of concentrations. When

$$d \ln a_i \simeq d \ln m_i = d\, m_i/m_i = (m_i'' - m_i')\, dx/m_i,$$

equation 6.4 becomes

$$-\mathbf{E}_1 = \Sigma(RT/\mathbf{F}) \int_0^1 (u_i/z_i)(m_i'' - m_1')\, dx/(B_1 + B_2\, x)$$

In this equation, the summation is over all of the *i*th ions in the solutions. This is a sum of constants times

$$\int_0^1 dx/(B_1 + B_2\, x)$$

The value of each integral is $(1/B_2) \ln [(B_1 + B_2)/B_1]$. Thus,

$$-\mathbf{E}_1 = \Sigma(RT/\mathbf{F})(u_i(m_i'' - m_i')/z_i\, B_2) \ln [(B_1 + B_2)/B_1]$$

When the values of B_1 and B_2 are inserted,

$$\mathbf{E}_1 = [(RT/\mathbf{F})\Sigma(u_i/z_i)(m_i'' - m_i')/\Sigma(m_j'' - m_j')\, u_j] \ln (\Sigma m_j'\, u_j/\Sigma m_j''\, u_j)$$

Since now summations are being made over all ions after integrations for each ion, i, the distinction between ions i and j is no longer useful. Thus the equation can be written in the manner shown in equation 6.6.

7

Transport Across Membranes

In this chapter on transport across membranes, emphasis will be placed upon general mechanisms of transport and on the basic physics and physical chemistry underlying these general mechanisms. Transport can be passive, in which case the solute moves down a concentration gradient or a potential gradient, or active, in which case free energy derived from the hydrolysis of adenosine triphosphate (ATP) or, in a few instances, some other phosphorylating agent drives solutes against gradients. Because of the enormous role played by the hydrolysis of ATP, a section of the thermodynamics of this reaction is included. Transport can be directly through the lipid bilayer, through pores or channels in the membrane, or by interaction with enzyme-like permeases imbedded in the membrane. Included is a discussion of anomalous osmosis, a phenomenon that has not thus far been demonstrated to play an important role but that has intriguing possibilities. No attempt is made in this chapter to deal with the enormous number of specific transport mechanisms uncovered by membrane biochemists and cell biologists, although a few examples are mentioned to illustrate principles.

ANOMALOUS OSMOSIS

Anomalous osmosis, discovered in the middle of the 19th century by Dutrochet, is defined as the movement of solvent across a membrane separating electrolyte solutions at different electrolyte concentrations at a rate that differs from that experienced in normal osmosis, either too fast or too slow, even in the wrong direction. When the flow is too fast, the anomalous osmosis is positive, and when it is too slow or in the wrong direction, it is negative. Physiologists have debated the role of anomalous osmosis in physiological processes since the discovery of the phenomenon. The consensus on whether or not a role exists has fluctuated. The current view is essentially negative. Be that as it may, the phenomenon is real; transport across artificial membranes does occur by this means. Sollner and collaborators have demonstrated the existence of both positive and negative anomalous osmosis with artificial membranes and have delineated conditions necessary for obtaining the phenomenon [1]. Among the necessary conditions are the following:

1. The membrane separating two electrolyte solutions must have charged pores of at least two kinds. The reason for this will become apparent in the discussion that follows.
2. The electrolyte must be at different concentrations on the two sides of the membrane.
3. The electrolyte must dissociate into positive and negative ions of distinctly different mobilities.
4. A potential drop must exist across the membrane.

All explanations of anomalous osmosis involve electroosmosis across the membrane. As was pointed out in Chapter 5, electroosmosis is the flow of a solution through capillaries or pores with charged surfaces under the influence of a potential gradient. The rate of flow depends not only on the potential gradient but also on the electrokinetic potential at the surface, which, in turn, depends upon the charge density of the pore surface and the ionic strength of the electrolyte solution. These statements are supported by equations 5.12, 5.28, and 4.15. Equation 5.12 relates electroosmotic mobility, u, defined as velocity per unit potential gradient, to the electrokinetic potential, ζ. $u = -\mathbf{D}\zeta/4\pi\eta$. Equation 5.28 can be written in the form, $\zeta = 4\pi\sigma/\mathbf{D}\boldsymbol{\kappa}$. Equation 4.15 shows the relationship between $\boldsymbol{\kappa}$ and ionic strength, μ. For an aqueous solution at 25°C, $\boldsymbol{\kappa} = 0.327 \times 10^8 \sqrt{\mu}$. All of these equations apply to limiting conditions that probably are not valid for pores in a membrane, but they do, nevertheless, illuminate the basic principles and show the important variables. Since, under conditions where electroosmosis can occur, charges of sign opposite those of the surface are in excess in the liquid, electroosmosis is a charge-carrying process. Thus, since anomalous osmosis depends on electroosmosis, it can occur only when there is a closed circuit across the membrane. Charges must flow in both directions between the solutions separated by the membrane. That is why it is necessary to have pores in the membrane of two different types, one type in which electroosmosis occurs and another type in which electric current corresponding to the excess charge in the liquid passing through the pores by electroosmosis can return. Anomalous osmosis in artificial membranes is a transient, nonequilibrium phenomenon. Since it depends on unequal concentrations of an electrolyte on the two sides of the membrane, it ultimately comes to a halt when equilibrium is approached. However, if, as in biological systems, there is an active process to maintain the electrolyte gradient, anomalous osmosis could be sustained indefinitely.

Many explanations of exactly how anomalous osmosis takes place can be found in the literature. Figure 7.1 illustrates an explanation that seems plausible to the author. The top of the figure illustrates the situation that would exist if three dry cells were arranged in the manner shown. The circle with the arrows shows the direction of flow of positive current, to the right above and to the left below. The positive current would actually flow backward through the lower dry cell. The potential is shown on the extreme right. The middle and bottom sections of the figure illustrate positive and negative anomalous osmosis, respectively. In both cases, a membrane with two kinds of charged pore separates hydrochloric acid solutions at a higher concentration c' on the left and a lower concentration c'' on the right. The solid arrows indicate the direction of flow of liquid associated with electroosmosis. Flow of liquid by ordinary osmosis is from right to left, from dilute HCl solution to concentrated HCl solution, illustrated by the dashed arrows. In both the middle and bottom portions of the diagram there are two kinds of charged pore, a narrow pore (N) above and a wide pore (W) below. The narrow pore has a negative charge and is therefore assumed to be permeable to hydrogen ions only. The potential developed across such a pore is given by equation 6.10. It is positive on the right. In both cases, potentials will also be developed across the wide pores (W). These pores are assumed to be permeable to both positive and negative ions, and therefore the potential will be given by equation 6.9. Thus the positive potential on the right will be higher in the upper pore than in the lower pore, just as in the illustration with the dry cells at the top of the diagram, and positive electricity will flow to the right above and to the left below as illustrated with the circles. Also, as shown by the equations on the right of the figure, the net potential across the membrane, **V**, will be positive on the right. If, as illustrated in the middle, the wide pores, (W), have a negative charge, the counter ions will have a positive charge, and electroosmosis will take place in the

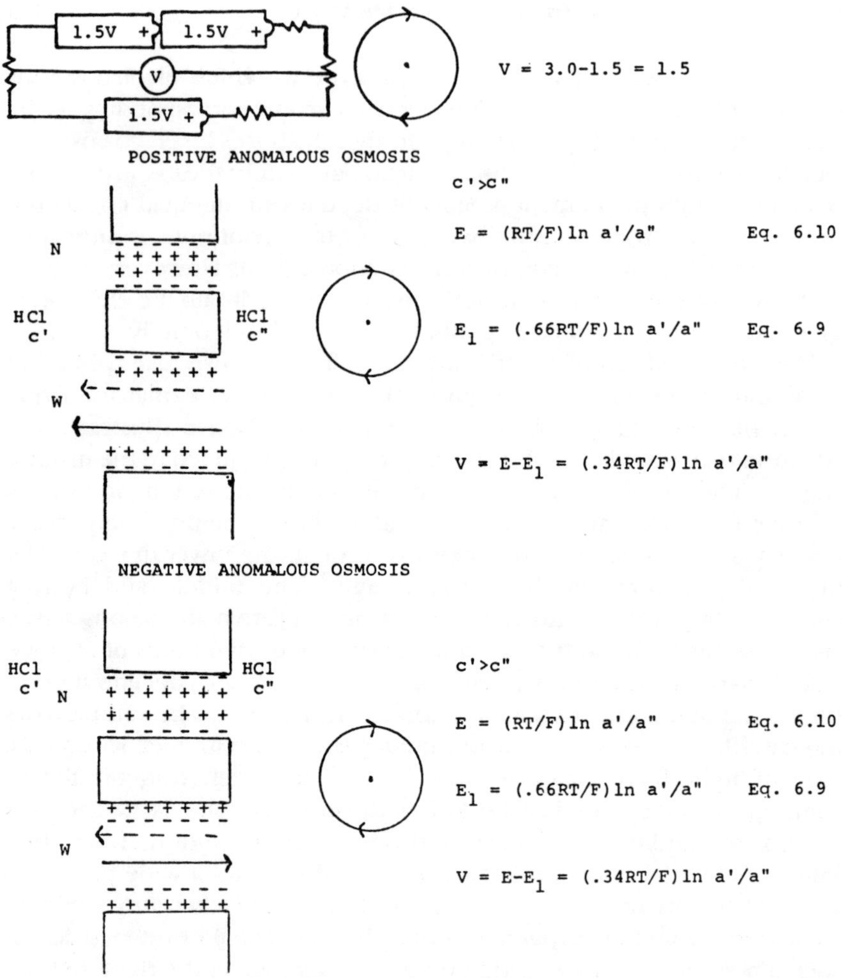

Fig. 7.1. Scheme illustrating the mechanism of anomalous osmosis. The illustration at the top shows how a positive current can flow against an opposing positive potential. The middle and bottom sections represent possible membrane structures that would show anomalous osmosis, positive for the middle structure and negative for the bottom structure. In both cases, the electrolyte concentration on the left, c′, is greater than that on the right, c″. In both cases, there are small pores through the membrane permeable only to positive ions. For both positive and negative anomalous osmosis, there are larger pores permeable to all ions. The difference is that for positive anomalous osmosis the large pore has the same sign as the small one, but for negative anomalous osmosis the large pore has charge of the opposite sign. The circles show the direction of circulation of positive charges. In both cases, the positive potential generated at the small pore is greater than that at the large pore, so that a situation comparable to the arrangement of the batteries at the top is realized. Dotted arrows show the direction of normal osmosis and solid arrows the direction of electroosmosis.

same direction as ordinary osmosis. The total rate will be greater than that of ordinary osmosis. If, on the other hand, the wide pores, (W), are positively charged, the counterions in the liquid will be negative, and positive current will flow from right to left by the device of negative counter charges moving from left to right. That is, electroosmosis will be from left to right, in the direction opposite that of ordinary osmosis. If the electroosmosis is strong enough, it will completely overbalance ordinary osmosis, and liquid will flow in the direction of concentrated electrolyte to dilute electrolyte.

It is worthwhile to ponder the possible role of anomalous osmosis, either positive or negative, for the transport of solutes, both charged and uncharged, across a membrane, even against a concentration gradient. Anomalous osmosis involves the movement by electroosmosis of solvent and everything dissolved in it except particles too large to pass through the pore. Thus charged or uncharged solute molecules move with the electroosmotic stream. With membranes permeable to water, the solvent can be forced back through the membrane by the pressures that build up inside the cell as the result of flow of liquid into the cell. Solutes can thus be concentrated on the side of the membrane toward which the anomalous osmotic flow occurs. In biology, anomalous osmosis is an idea whose time has not come. Whether or not it will ever come remains to be seen. For the present, it should be held in reserve.

DIFFUSION AND FACILITATED DIFFUSION

Some materials are transported across membranes by ordinary diffusion. The evidence seems to indicate that water is transported in this manner. The possibility exists that certain other materials soluble in the lipid bilayer could also be transported by diffusion. For a small area of membrane that can be considered, for simplicity, to be flat, equation 3.3 can be integrated to yield

$$(\Delta S/\Delta t = -DQ/L)(c_i - c_o) \qquad (7.1)$$

In this equation, c_i means the concentration of solute inside the cell and c_o is that outside. When c_i is much smaller than c_o,

$$\Delta S/\Delta t = (DQ/L)\, c_o \qquad (7.2)$$

While Q is the area and L the thickness, in the case of membranes the total area might not be permeable; Q/L is thus merely a membrane constant.

Many, perhaps most solutes can pass through a biological membrane only with the aid of carrier molecules in the membrane. These carriers behave somewhat like enzymes; they combine with solute outside the membrane and dissociate the solute inside. Define $\mathbf{n}$ as the number of carrier molecules per unit area of membrane, $\mathbf{n}_c$ the number combined with solute, and $\mathbf{n}_f$ the number uncombined. The number of moles of solute combining in unit time with carriers on the outside is $k_1 c_o \mathbf{n}_f = k_1 c_o(\mathbf{n} - \mathbf{n}_c)$. The number dissociating on the outside is $k_{-1}\mathbf{n}_c$. Similarly, the number of moles combining on the inside is $k_{-2}\, c_i\, (\mathbf{n} - \mathbf{n}_c)$, and the number of moles dissociating per unit time and per unit area on the inside is $k_2\mathbf{n}_c$. At steady state, when $\mathbf{n}_c$ is a constant,

$$k_1\, c_o\, (\mathbf{n} - \mathbf{n}_c) + k_{-2}\, c_i\, (\mathbf{n} - \mathbf{n}_c) = k_{-1}\, \mathbf{n}_c + k_2\, \mathbf{n}_c \tag{7.3}$$

This can be rearranged to yield $(k_1\, c_o + k_{-2}\, c_i)\, [(\mathbf{n} - \mathbf{n}_c)/\mathbf{n}_c] = k_2 + k_{-1}$. If it is assumed that the rate constant for combining of carrier molecules with solute is the same on the inside and on the outside, that is, if $k_1 = k_{-2}$, this equation can be rearranged to the form,

$$(c_o + c_i)\, [(\mathbf{n} - \mathbf{n}_c)/\mathbf{n}_c] = (k_2 + k_{-1})/k_1 \equiv K_m \tag{7.4}$$

When equation 7.4 is solved for $\mathbf{n}_c$,

$$\mathbf{n}_c = (c_o + c_i)\, \mathbf{n}/(K_m + c_o + c_i)$$

and when c_i is negligibly small compared to c_o,

$$\mathbf{n}_c = c_o\, \mathbf{n}/(K_m + c_o) \tag{7.5}$$

Under the condition that c_i is negligibly small, the rate of transport, v, from outside to inside is given by $v = k_2\, \mathbf{n}_c$. The maximum rate of transport will occur when all of the carrier molecules are combined, that is, when $v_{max} = k_2\, \mathbf{n}$. Thus,

$$v = v_{max}\, c_o/(K_m + c_o) \tag{7.6}$$

In this equation, v corresponds to $\Delta S/\Delta t$ of equation 7.2. Equation 7.6 is of exactly the same form as the basic equation for enzyme kinetics. Since the assumptions involved in deriving it are like those involved in developing the Michaelis-Menten equation of enzyme kinetics, K_m can be called a Michaelis-Menten constant. The reciprocal of equation 7.6 is

$$1/v = K_m/(v_{max}\, c_o) + 1/v_{max} \qquad (7.7)$$

Equation 7.6 shows that when c_o is very small, the rate of transport is directly proportional to c_o just as it is in the case of simple diffusion, corresponding to equation 7.2. However, when c_o is very large, the rate of transport is independent of c_o, unlike the case of simple diffusion, and is equal to v_{max}. Carrier-mediated transport can be analyzed by the use of equation 7.7, just as in the case of enzyme kinetics. Figure 7.2a illustrates such a plot. However, as illustrated by Figure 7.2b, reciprocal plots of experimental data do not always yield a straight line, as required by equation 7.7; rather, at high values of c_o when $1/c_o$ approaches zero, 1/v also approaches zero. At intermediate and large values of $1/c_o$ the straight line is obtained. This can be interpreted to mean that in such a case the transport is partially carrier mediated and partially simple diffusion.

Transport of the sort described, whether simple diffusion or carrier-mediated, is passive. The free energy is given by

$$\Delta G = RT\ln c_i/c_o \qquad (7.8)$$

Equilibrium is established when $c_i = c_o$. If the solute is charged and if there is a potential across the membrane, then an electrical component must be added to the free energy equation

$$\Delta G = RT\ln c_i/c_o + z\mathbf{F}\Delta\mathbf{V} \qquad (7.9)$$

In this equation, z is the net charge on the solute, **F** is the faraday constant, and **ΔV** is the potential change across the membrane. In this case, equilibrium does not occur when $c_i = c_o$; the possibility is thus afforded for accumulating higher concentrations on one side or the other of the membrane. As long as the potential across the membrane is maintained, in part at least by metabolic processes, as discussed in

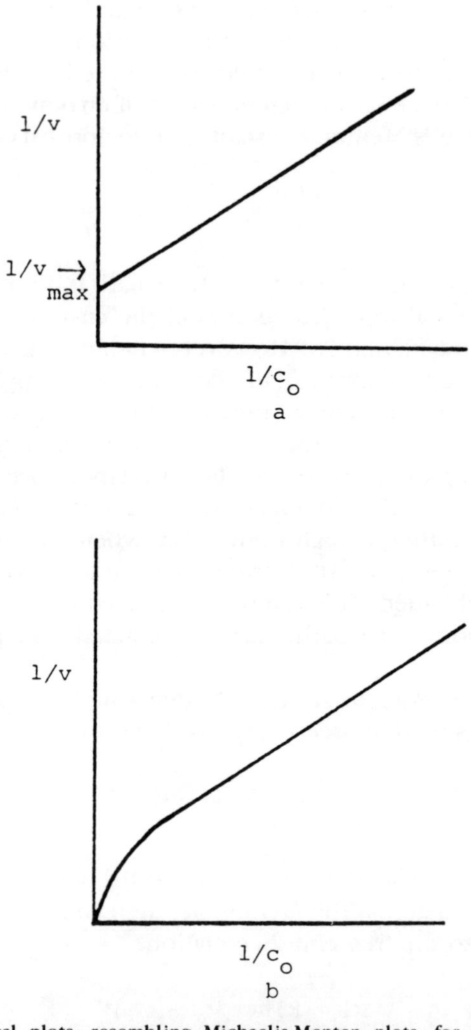

Fig. 7.2. Reciprocal plots, resembling Michaelis-Menten plots, for facilitated diffusion. **a:** Uncomplicated facilitated diffusion. **b:** Mixed simple diffusion and carrier-mediated diffusion.

Chapter 6, ions can move across the membrane even against a concentration gradient. A carrier protein in the membrane might be involved, but the energy is derived immediately from the potential gradient.

Exchange diffusion is carried out by the anion-transport protein of erythrocytes. In the lung capillaries, carbon dioxide in the form of the bicarbonate ion diffuses out of the erythrocyte through the transport protein in one-to-one exchange for chloride ions diffusing in. The flow of the two ions in opposite directions is determined energetically by the sum of the concentration gradients of the two ions. Such situations are called antitransport. In the human erythrocyte glucose is transported by carrier facilitated diffusion. The process is passive, that is, glucose flows down a concentration gradient. Carrier-mediated transport of glucose and of other solutes in other membranes can also be active, involving the hydrolysis of ATP. A discussion of the thermodynamics of ATP hydrolysis is presented in the next section.

An interesting transport mechanism, a symtransport mechanism involving chemiosmotic coupling, involves the simultaneous movement of one H^+ ion and one lactose molecule across the *E. coli* membrane, mediated by a lactose permease. Hydrogen ions flowing down a concentration gradient can carry with them lactose molecules against a concentration gradient. Both hydrogen ions and lactose molecules can be transported against concentration gradients in a potential gradient. Lactose moving down a concentration gradient can move hydrogen ions against a concentration gradient. While the maintenance of the driving gradients is undoubtedly active, the actual transport process itself is passive. Situations like this are sometimes called secondary active transport.

THERMODYNAMICS OF ATP HYDROLYSIS

The ultimate and penultimate phosphate groups of adenosine triphosphate (ATP) are attached to their neighboring phosphate groups by what biochemists call high-energy phosphate bonds. Bull [2] pointed out that this choice of name is unfortunate. Physical chemists interpret the strength of a chemical bond in terms of the enthalpy decrease when that bond is formed or the *heat required to break* that bond. High-energy phosphate bonds are almost the opposite; they are bonds which upon hydrolysis yield a large *decrease in free energy.* Bull suggested that a more suitable term would be hypergonic bonds. Be that as it may, these phosphate bonds play an enormous role in metabolism. The most important of all of the compounds containing high-energy phosphate bonds is ATP, for when the hydrolysis of ATP is coupled with reactions for which the free energy change is positive, the net or overall free energy change is negative and the coupled reaction is spontaneous.

The hydrolysis reaction can be represented by the formula, ATP + $H_2O \rightleftarrows$ ADP + P_i. This reaction, however, is complicated by two considerations. First, all of the phosphate compounds dissociate hydrogen ions, the numbers of which depend upon the pH. Second, all are capable of binding divalent cations, of which Mg^{2+} can be taken as a representative. Phillips et al. [3] have considered in detail the various equilibria involved.

The standard free energy, $\Delta G°$, for this reaction would, following the normal conventions, be the free energy when all reactants and products were at unit activity. A more convenient standard free energy is that at which reactants and products are at unit activity but at which the activity of the hydrogen ion, a product of the reaction under normal physiological conditions, is that at pH 7.0. This standard free energy is designated as $\Delta G^{o\prime}$. The equilibrium of the reaction as written above at pH 7 is so far to the right that determination of the equilibrium constant directly would require analytical precision greater than anything available. Thus, $\Delta G^{o\prime}$ was determined indirectly by two reactions. Levintow and Meister [4] determined the equilibrium constant at pH 7, 0.2 ionic strength, 35 mM Mg^{2+} and 37°C to be 1,200 ± 45 for the reaction glutamate + ATP + ammonia $\rightleftarrows$ glutamine + ADP + P_i. This equilibrium constant can be designated K_1. Benzinger et al. [5] determined the equilibrium constant, K_2, under approximately the same conditions for the reaction glutamine + $H_2O \rightleftarrows$ glutimate + ammonia. K_2 was evaluated to be 225 ± 25 calories per mole. These two reactions add up to ATP + $H_2O \rightleftarrows$ ADP + P_i, and the equilibrium constant, K, equals K_1 K_2 or 270,000. Many refinements and corrections were discussed by Phillips et al. [3] in interpreting these data. For the most part, these contribute no great change in the final result; for the sake of simplicity they will be ignored in the present treatment.

$\Delta G^{o\prime}_{37} = -RT \ln K = -7{,}700$ cal/mole. The enthalpy, ΔH^o, for the hydrolysis of ATP at pH 7 was determined by Kitzinger and Benzinger [6] and by Podolsky and Morales [7] to be −4.8 kcal and −4.7 kcal, respectively. The mean value, −4.75 kcal/mole, is taken for ΔH^o. From these values of $\Delta G^{o\prime}$ and ΔH^o, using the equation $\Delta G = \Delta H - T\Delta S$, ΔS can be calculated to be +9.52 eu. With this value for the entropy and the enthalpy, $\Delta G^{o\prime}_{25}$ can be calculated to be −7.6 kcal per mole. This is the value at pH 7, 0.2 ionic strength, and 35 mM Mg^{2+}. The actual value of $-\Delta G$ can be much higher than this.

$$\Delta G = \Delta G^{o'} + RT \ln [ADP][P_i]/[ATP] \quad (7.10)$$

In human erythrocytes the concentrations of P_i, ADP, and ATP can be taken to be 1.65, 0.25, and 2.25 mM, respectively. When these values are substituted, ΔG is -12.7 kcal/mole. Since ΔH is -4.75 kcal/mole, $-T\Delta S$ is -7.9 kcal/mole. Thus it can be seen that the hydrolysis of ATP is driven both by a decrease in enthalpy and by an increase in entropy. Under the conditions just mentioned, the entropic contribution is nearly twice the enthalpic contribution; the reaction is largely entropy-driven. The calculations of Phillips et al. [3] show that the free energy decrease is about 25% higher at pH 9 than at pH 6 and that it is somewhat lower in the presence of magnesium ion than in its absence. Thus the actual value of ΔG in a biological or biochemical reaction can be somewhat different from the value calculated here.

Under physiological conditions, ATP is stable even though the equilibrium is far on the side of ADP + P_i. Only when a specific enzyme is present do reactions such as that between glutamate, ATP, and ammonia take place. The specific enzyme used in the studies referred to above was glutamyl transferase from green peas. A typical action of ATP in metabolism is the transfer of a phosphate group from ATP to D-glucose, catalyzed by the enzyme hexokinase.

$$\text{ATP} + \text{D-glucose} \overset{Mg^{2+}}{\rightleftarrows} \text{ADP} + \text{D-glucose 6-phosphate}$$

Glucose 6-phosphate has a higher energy content than glucose and can now undergo further enzymatic reactions. ATP can also serve as a source of free energy decrease to transfer electrolytes and other solutes against concentration gradients or to break the bonds between protein molecules such as that between actin and myosin heads in actomyosin.

ACTIVE TRANSPORT

There are many examples of active transport across membranes involving ATP or some other source of energy. The sarcoplasmic reticulum of muscle cells contains a calcium ATPase that catalyzes the reaction by which calcium ions are concentrated in the reticular lumen. The total reaction involves

$$2\,Ca^{2+}\ \text{(outside)} + \text{ATP (outside)} \overset{Mg^{2+}}{\rightleftarrows} \text{ADP (outside)} + P_i\ \text{(outside)} + 2\,Ca^{2+}\ \text{(inside)}$$

Thus the calcium ion moves against its concentration gradient with the hydrolysis of ATP as the source of energy. In the cytoplasmic membrane of the epithelial cells lining the stomach, there is a H^+-translocating ATPase. This system drives H^+ ions against its concentration gradient into the stomach in a manner comparable with that described above.

Bacterial, chloroplast, and mitochondrial membranes contain a system involved in the synthesis of ATP. This system involves a channel for H^+ ions called F_0 and associated with it on the inside an ATPase called F_1. When hydrogen ions flow down a concentration gradient from outside to inside, the energy is provided for the enzymatic synthesis of ATP from ADP + P_i inside. This same system can operate in reverse to force hydrogen ions across the membrane against a concentration gradient, with the energy supplied by the hydrolysis of ATP inside.

Finally, some materials are transported across a membrane by the action of ATP, accompanied by chemical modification of the substrate. This is called group translocation. In many bacteria, sugars taken up from the surrounding medium are converted into sugar monophosphates as they are transported through the membrane. The permease for the sugar is also an enzyme that catalyzes both the phosphorylation and the transport of the substrate.

MECHANISMS OF PERMEASE FUNCTION

The question of how permeases act in assisting transport across biological membrane is under active investigation. Few final answers are available. However, sufficient information has been obtained to justify consideration of some general models, models that in some cases are probably correct.

Some permeases are mobile carriers, molecules that engulf hydrophilic solutes and shuttle through the hydrophobic membrane. Evidence for this kind of mechanism comes from the observation that transport occurs only at higher temperatures, temperatures above the melting point of the lipid membrane, and not at lower temperatures, where the membrane is crystalline. An example is the potassium-transporting antibiotic, valinomycin. This is a cyclical peptide extremely selective for binding K^+ ions. In an aqueous medium, the hydrophilic side chains are on the outside of the ring. When a K^+ ion is encountered, a conformational rearrangement takes place, such that the ion is bound on the interior by complexing with hydrophilic side chains, while the hydrophobic side chains of the peptide are on the outside. This renders the complex soluble in lipid bilayers, making both biological membranes and artificial bilayers permeable to potassium. When the complex reaches the aqueous environment on the other side of the membrane, it again undergoes a conformation change to release the K^+ ion. On the side of the membrane with the higher K^+ ion concentration, more ions are bound than released, while on the side with the lower concentration, the opposite is true.

Usually the facilitators are proteins that span the membrane. Models usually represent these proteins as oligomers, primarily dimers, which form pores or channels through the membrane. In some cases, the pores are nonselective; they allow many different solutes to permeate, as long as they are small enough to pass through. An example is the permeability attributable to a group of proteins called polins found in the outer membrane of *E. coli.* These molecules have been shown by electron-microscopy to be arranged in a hexagonal lattice of trimers. They form pores with a diameter of about 1 nm, large enough to allow molecules with molecular weights below 600 to pass. When polin molecules are incorporated into artificial phospholipid vesicles, they allow rapid diffusion of most small hydrophilic molecules through the bilayer.

Many channels, however, have great specificity. For those involved in passive transport it is necessary only that the permease bind a specific solute from solution on one side of the membrane and release it on the other. A simple equilibrium between bound and unbound solute is sufficient to explain the operation in this case. On the side of the membrane adjacent to the higher concentration of the solute, the net reaction will be in the direction of binding, while on the side of the membrane in contact with the lower concentration of solute, the net reaction will be in the direction of dissociation.

The situation is more complicated in the case of active transport, where energy must be supplied to transport solutes against concentration or potential gradients. Suppose that the permease is a protein dimer spanning the membrane that is capable of undergoing a change in conformation. On the side of the membrane adjacent to the low concentration of specific solute, the conformation is such that a strong binding site for the solute is presented. When the solute is bound, phosphorylation of the permease, usually by the hydrolysis of ATP but sometimes by the hydrolysis of other phosphorylating agents, causes a conformational change such that the pore is closed on the low concentration side and opened on the high concentration side; simultaneously the strength of the binding site is greatly reduced. Dissociation will then occur on the high concentration side, accompanied by dephosphorylation and reversion to the original conformation. Examples are known in which there is a one-to-one correspondence between the number of ATP molecules hydrolyzed and the number of solute molecules transported, but ratios of one to two or one to three are also known. Some of these systems can involve two-way transport, one solute in and the other out. An example is the $Na^+ - K^+$ ATPase described in Chapter 6 in which sodium and potassium ions are both transported against concentration gradients, one from outside to inside and the other from inside to outside.

Klingenberg [8] has made a thoughtful analysis of the probable arrangement of permease oligomers in the membrane. He reviewed studies made on natural membranes and on isolated carrier proteins. Two permease subunits span the membrane with their twofold axes of symmetry oriented perpendicular to the membrane and with a channel between them. The hydrophobic surfaces are in contact with the surrounding lipid, and specific binding sites for solutes and for phosphorylation are in the channel between the subunits. Figure 7.3 illustrates how a gated pore might operate.

TRANSPORT OF MACROMOLECULES AND PARTICLES

In the usual case, macromolecules are transported across membranes of the cell by exocytosis or endocytosis, to be discussed in the next paragraph. However, in some instances large molecules pass directly through the membrane. The genetic constitution of some bacteria can be changed by exposure to purified DNA, which crosses both the cell wall and plasma membrane. Some proteins are directly transferred across the membrane of the endoplasmic reticulum in eucaryotic cells. Bacteriotoxins are also able to penetrate animal cell membranes. The mechanisms of these transports are not fully understood, but specific channels might be involved.

Transport of macromolecules and even large particles across cell membranes normally takes place by exocytosis or endocytosis. In exocytosis, cells surround the macromolecules in intracellular vesicles, which then fuse with the plasma membrane and open to the outside. Thus the macromolecule is extruded from the cell. Cells take in macromolecules and even particles as large as bacteria by the reverse process, endocytosis. When small amounts of fluid and its dissolved components are

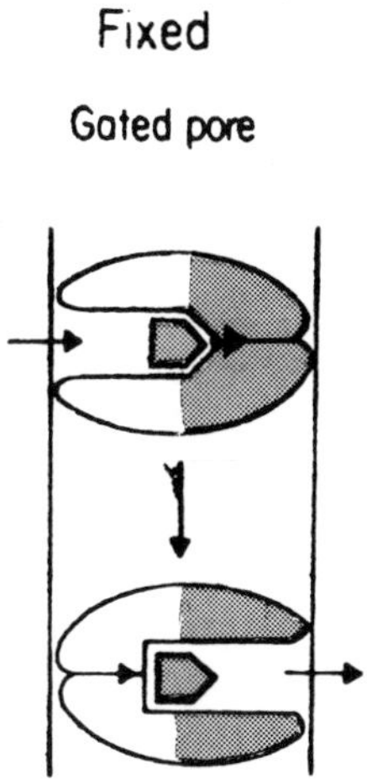

Fig. 7.3. Schematic illustration of how a fixed gated pore might operate. The permease consists of two subunits spanning a membrane with a channel between them. Solute enters from one side and is bound, causing conformational changes that open the channel to the opposite side. If the binding energy is lowered as a result of the conformation change, with energy supplied by ATP hydrolysis, the solute can be released against a concentration gradient. Reprinted from Klingenberg [8] with permission of the copyright owner, Nature.

ingested, this is called pinocytosis or cell drinking. When large particles such as microorganisms are ingested, the process is called phagocytosis, or cell eating. In both cases, an invagination of the cell membrane occurs. The foreign macromolecule or particle or the external liquid is thus enclosed. As the process proceeds, the membrane fuses external to the ingested membrane-surrounded material. Finally, the surrounding membrane is removed, and the material is released inside the cell. It is generally assumed that endocytosis and exocytosis are active processes involving the expenditure of energy, possibly derived from the hydrolysis of ATP.

REFERENCES

1. Sollner, K. and Grollman, A. Zeitsch f. Elektrochemie *38:*274, 1932.
2. Bull, H.B. Introduction to Physical Biochemistry, Edition II. Philadelphia: F. A. Davis Co., 1971.
3. Phillips, R.C., George, P., and Rutman, R.J. J. Biol. Chem. *244:*3330–3342, 1969.
4. Levintow, L. and Meister, A. J. Biol. Chem. *209:*265, 1954.

5. Benzinger, T., Kitzinger, C., Hems, R., and Burton, K. Biochem. J. *71:*400, 1959.
6. Kitzinger, C. and Benzinger, T. Z. Naturforsch. *10:*375, 1955.
7. Podolsky, R.J. and Morales, M.F. J. Biol. Chem. *218:*945, 1956.
8. Klingenberg, M. Nature *290:*449–454, 1981.

8

Entropy-Driven Processes in Biology. I. Mechanism and Significance

Spontaneous entropy-driven processes are those that occur primarily because of an increase in entropy. Entropy-driven assemblies of microfilaments and microtubles and other structures important in the functioning of living cells are frequently found. The present chapter deals primarily with the basic mechanism of entropy-driven assemblies and a discussion of the significance of such processes. Much of our understanding of the factors that influence or control entropy-driven processes derives from extensive studies on the polymerization of tobacco mosaic virus protein carried out in the author's laboratory over a three-decade period. The then current status of this research, supplemented by discussions of other important entropy-driven processes in biology that had come to the attention of the author, was published in 1975 [1]. Considerable progress in understanding such processes has since occurred. The most important of the earlier studies will be reviewed briefly, and the more recent studies that have led to an understanding of the significance of entropy-driven assembly processes will be discussed in greater detail. In the next chapter, entropy-driven processes in living organisms, particularly those processes that affect cellular shape and movement, will be treated.

THERMODYNAMIC BACKGROUND

To make it perfectly clear what is meant by an entropy-driven process, it is helpful to review a few aspects of thermodynamics. The first law can be represented symbolically by equation 8.1:

$$\Delta E = \mathbf{Q} - W \qquad (8.1)$$

E is the total or the internal energy; **Q** is the heat absorbed, and W is the work done. For a reversible process,

$$dE = dQ - dW \tag{8.2}$$

In this equation, dW is the sum of the work done by expansion against the atmosphere, the mechanical work, PdV, and the nonmechanical work, such as electrical work, dW_{nm}.

$$dE = dQ - PdV - dW_{nm} \tag{8.3}$$

When there is no nonmechanical work, that is, when dW_{nm} is 0, and when volume is kept constant, that is, when dV is 0,

$$dE = (dQ)_V \quad ; \quad \Delta E = Q_V \tag{8.4}$$

ΔE can be evaluated by measuring the heat absorbed or evolved when a reaction is carried out in a bomb calorimeter.

Enthalpy, H, is defined by equation 8.5:

$$H \equiv E + PV \tag{8.5}$$

$$dH = dE + PdV + VdP \tag{8.6}$$

By substituting equation 8.3 for the case in which there is no nonmechanical work for dE,

$$dH = dQ + VdP \tag{8.7}$$

For processes carried out at constant pressure, $dP = 0$,

$$dH = (dQ)_P \quad ; \quad \Delta H = Q_P \tag{8.8}$$

ΔH can be determined for a process or a reaction carried out at constant pressure by measuring the heat absorbed or evolved with a calorimeter. Because biological processes normally occur at constant pressure, equations derived for the condition of constant pressure are directly applicable. For processes carried out in solution, except those involving the adsorption or the evolution of gas, dV is either 0 or very small. Thus, for most biological processes, it follows from equation 8.6 that $dH \simeq dE$ and $\Delta H \simeq \Delta E$.

When chemical bonds, either strong or weak, are formed, heat is evolved, and when they are broken, heat is absorbed. The enthalpy change, ΔH, is the sum of the heat absorbed in breaking old bonds and the heat evolved (negative) in forming new bonds. In most chemical reactions involving the formation of large structures from smaller molecules, heat is evolved because the net process is the formation of new bonds; ΔH is negative. Such processes are exothermic.

Enthalpy, H, is fractionated into two terms. $H = G + TS$ or, at constant temperature, T, $\Delta H = \Delta G + T\Delta S$. ΔG, the change in Gibbs free energy, is the portion of ΔH that is free to do work, such as electrical work, for example, and ΔS is the increase in entropy.

$$\Delta G = \Delta H - T\Delta S \tag{8.9}$$

Equation 8.9 is a deduction from the second law of thermodynamics. Further deductions show that, when there is no nonmechanical work, ΔG is 0 for isothermal reversible processes (equilibrium processes) and negative for all such spontaneous processes. In the usual chemical synthesis of large molecules from smaller ones, ΔS is negative because there is an increase in order; ΔG is negative because ΔH is negative. However, when an assembly process is endothermic, that is, when ΔH is positive, ΔG can be negative only when ΔS is positive and TΔS exceeds ΔH in magnitude. Such processes are defined as entropy-driven reactions.

When reactants and products are at unit activity, equation 8.9 takes the form $\Delta G^\circ = \Delta H^\circ - T\Delta S^\circ$. The superscript is critically important for G and S. However, because H per mole of reactant varies only slightly with concentration for most biologically important macromolecules, the superscript for H can be omitted without significant conceptual error in the present context. When the expression for ΔG° given by equation 8.10, which was derived from thermodynamic considerations, is substituted into equation 8.9 written in the form corresponding to the standard state, equation 8.11 is obtained. K_a is the equilibrium constant on the activity basis.

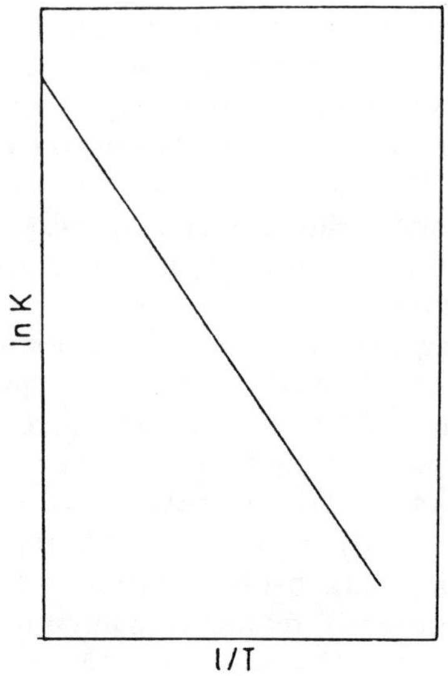

Fig. 8.1. Theoretical plot of the natural logarithm of the equilibrium constant, ln K, against the reciprocal of the temperature in Kelvin, 1/T. From Lauffer [1], reprinted with permission of the copyright owner, Springer Verlag.

$$\Delta G^\circ = -RT \ln K_a \tag{8.10}$$

$$\ln K_a = \Delta S^\circ/R - \Delta H^\circ/RT \tag{8.11}$$

When ΔH° is positive, as it is in endothermic and, therefore, entropy-driven processes, a plot of $\ln K_a$ versus l/T yields a straight line with intercept, $\Delta S^\circ/R$, and negative slope, $\Delta H^\circ/R$. This is illustrated in Figure 8.1. Thus graphs such as those illustrated in Figure 8.1 afford a convenient and foolproof method of identifying entropy-driven processes; when ΔH° is positive, the process is of necessity entropy-driven because only when ΔS° is positive can, by equation 8.9, ΔG° be negative, and ΔG° must be negative if the reaction is to proceed.

Historically, some important entropy-driven biological processes have been discovered by studying the effect of hydrostatic pressure, P, on the equilibrium constant, K_a. In accordance with the Le Chatelier principle, when there is an increase in volume, even a very small one, in a reaction, the process can be reversed by the application of pressure. From thermodynamics, when temperature is constant,

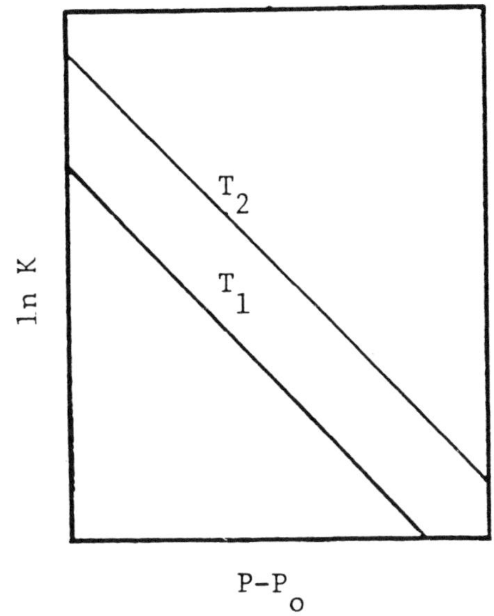

Fig. 8.2. Theoretical plot of the natural logarithm of the equilibrium constant, ln K, against pressure, $P - P_o$. Plots are shown for a higher temperature, T_2, and the lower temperature, T_1. From Lauffer [1], reprinted with permission of the copyright owner, Springer Verlag.

$$\partial\bar{G}°/\partial P = \bar{V} \quad ; \quad \partial\Delta\bar{G}°/\partial P = \Delta\bar{V} \tag{8.12}$$

When the value for $\Delta G°$ given by equation 8.10 is substituted and integration is performed, equation 8.13 results.

$$\ln K_a = \ln K_{ao} - (P - P_o)\,\Delta\bar{V}/RT \tag{8.13}$$

In these equations, the bar above a symbol signifies that the parameter applies to one mole of some participant in the reaction and the subscript, o, means the value at atmospheric pressure. When $\ln K_a$ is plotted against $(P - P_o)$ at constant values of T, straight lines with negative slopes, as illustrated in Figure 8.2, are obtained. The slope is $\Delta\bar{V}/RT$, and the intercept is $\ln K_{ao}$. $\Delta\bar{V}$ is positive for most biological assemblies, whether entropy-driven or enthalpy-driven. However, entropy-driven processes can be identified by the fact that for these processes alone, $\ln K_a$ increases with temperature, as is illustrated in Figure 8.2. Since temperature on the Kelvin scale varies only slightly on

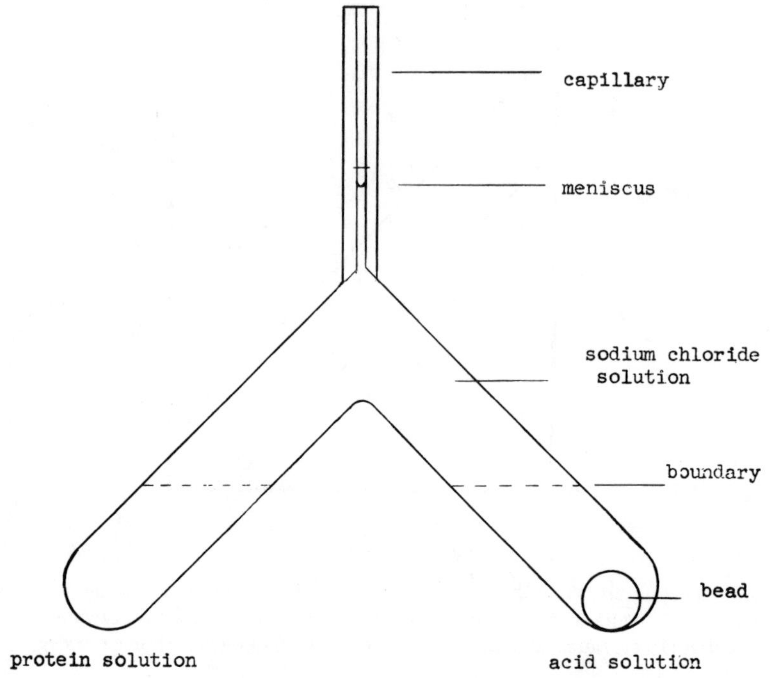

Fig. 8.3. Dilatometer for determining volume change accompanying polymerization of protein subunits. From Lauffer in Symposium on Foods; Proteins and Their Reactions, H.W. Schultz and A.F. Anglemier (eds), reprinted with the permission of the copyright owner, AVI Publishing Corp.

a fractional basis in the temperature range over which biological materials can be investigated, the slopes of graphs like those illustrated in Figure 8.2 are, for all practical purposes, parallel.

There is a slight increase in volume in assembly reactions because the partial specific volume changes slightly when subunits polymerize or assemble. This can be measured with a Linderstrøm-Lang dilatometer like that shown in Figure 8.3. If unpolymerized material, a protein, for example, is placed in one arm of the tube and a chemical that will cause polymerization, for example, an acid, is placed in the other and then the apparatus is carefully filled with solvent into the capillary, the change in partial specific volume can be determined by measuring the change in volume of the solution indicated by a slight change in the level in the capillary when the chemical and the protein are mixed. This volume change divided by the mass of protein in the vessel is the change in partial specific volume, ΔV. In this manner, Stevens and Lauffer [2] measured the increase in partial specific volume when tobacco mosaic virus protein subunits polymerize into helical rods to be 0.007 ml/gm.

POLYMERIZATION OF TMV PROTEIN

Because of its abundant availability, the coat protein of tobacco mosaic virus (TMV) has been the most important tool for the investigation of the nature of entropy-driven association of biologically important materials. Lauffer described the status of such investigations prior to 1975 in his book *Entropy-Driven Processes in Biology* [1]. In the current discussion, the most important aspects of the early work will be recapitulated briefly and the contributions of TMV protein polymerization to the understanding of entropy-driven biological processes made since that book was written will be reviewed in detail. The protein subunit, whose amino acid sequence is known, has a molecular weight of 17.53×10^3. It has a complex shape, but for the purposes of simplified physical reasoning, a prolate ellipsoidal revolution with a major semi-axis of 35 Å and a minor semi-axis of 12.25 Å is a reasonably satisfactory model. At a concentration of about 1 gm/l in 0.1 ionic strength (μ) buffers at pH values between 6.5 and 8 at temperatures of a few degrees C, the protein exists as an equilibrium mixture of the subunit and a trimer with molecular weight of 52.59×10^3 daltons. There is the possibility of other oligomers, but the trimer greatly predominates over all others, including the monomer. This equilibrium product is called A protein, considered for simplicity to be the trimer. Lauffer et al. [3] discovered in 1958 that TMV protein dissolved in a buffer at pH 6.5 could be polymerized into helical rods by raising the temperature to 25°C or 30°C and that the process could be reversed by lowering the temperature to 4°C or 5°C. Thus the polymerization of the protein was shown to be an endothermic and therefore entropy-driven process. It was also shown in this original study that at pH 5.0 the A protein exists only in the polymerized state and at pH 7.7 only in the unpolymerized state over the same temperature range. The authors recognized that the increase in entropy required to provide the necessary decrease in Gibbs free energy for such an endothermic reaction leading to the formation of a highly ordered structure could be understood only in terms of interaction between the protein and the solvent. They postulated that the overall reaction of polymerization

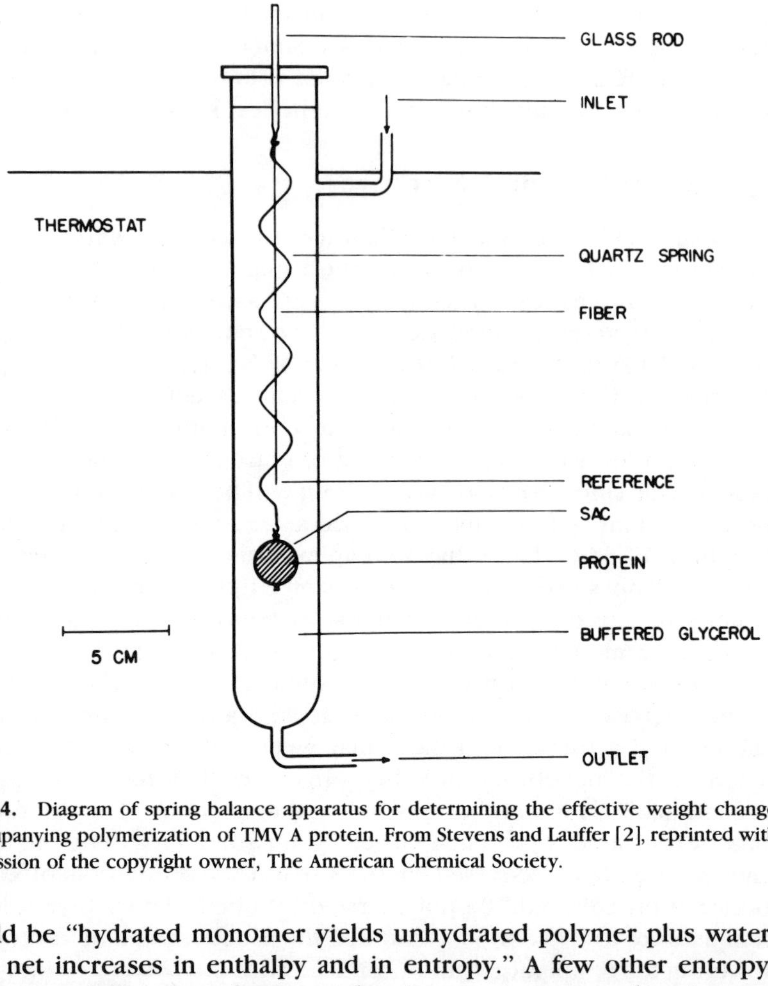

Fig. 8.4. Diagram of spring balance apparatus for determining the effective weight change accompanying polymerization of TMV A protein. From Stevens and Lauffer [2], reprinted with permission of the copyright owner, The American Chemical Society.

would be "hydrated monomer yields unhydrated polymer plus water, with net increases in enthalpy and in entropy." A few other entropy-driven associations were known at that time; a common explanation was release of solvent to provide the increase in entropy. However, this was purely theoretical speculation.

Stevens and Lauffer [2] used a spring balance, illustrated in Figure 8.4, to demonstrate that water molecules actually are released and to estimate how many. Detailed descriptions of the intricacies and complications involved in interpreting this experiment can be found in previous publications [1, 2]. The gist of the idea, stripped of all complications, is that if unpolymerized TMV protein dissolved in a buffered glycerol solution at pH 7.5 is placed inside the cellophane sac and is allowed to come to equilibrium with buffered glycerol in the external solution, what will be weighed by the spring balance is protein and water bound to protein, corrected for buoyancy. When the pH of the whole system inside and outside the sac is lowered to 5.5, the protein will polymerize. The released water will be freed to dialyze through the cellophane and come to equilibrium with the external solution; it will therefore no longer be weighed. Thus the change in weight of the contents of the sac can be determined by the change in extension of the previously calibrated quartz helix. Similar but more complicated experiments were later carried out by Jaenicke and Lauffer [4], in which polymerization was brought about by raising temperature. The results of the two experiments are in satisfactory agreement, yielding an average of 0.03 gm of water released per gm of protein polymerized. This corresponds to 29 water molecules per protein subunit or 87 per A-protein particle (trimer). The estimated uncertainty of this number is less than ± 10%. Thus the theory that the required increase in entropy derives primarily from release of water molecules for the polymerization of TMV A protein was substantiated by experiment. It is a reasonable inference that in other biologically important entropy-driven associations, which, in many respects, resemble the polymerization of TMV protein, the increase in entropy also comes primarily from the release of water molecules.

Early work outlined in a rough sort of way the stages in the polymerization of TMV A protein. It was Ansevin and Lauffer [5] who established in 1959 that an equilibrium existed between material sedimenting at 1.9 svedbergs and material sedimenting at 4 svedbergs, with the latter overwhelmingly the predominant form in solutions at a concentration of 1 gm per liter at pH values above 6.5. They identified the 1.9S component as the ultimate protein repeat unit in the TMV structure. Caspar [6] later identified 4S as a cyclical trimer and attempted to explain why it should be a stable intermediate. Banerjee and Lauffer [7] showed by osmotic pressure measurements that the very earliest stages of polymerization, covering the range of numbers of average molecular weight 50,000–75,000 daltons, could be explained in terms of an isodesmic mechanism, that is, it followed the mathematics of condensation polymerization. The idea was that A particles (trimers of the fundamental protein molecule) combine with A to form A_2, and A_2 combines either with A to form A_3 or with A_2 to form A_4. Durham and Klug [8] proposed another mechanism; A combines with protein

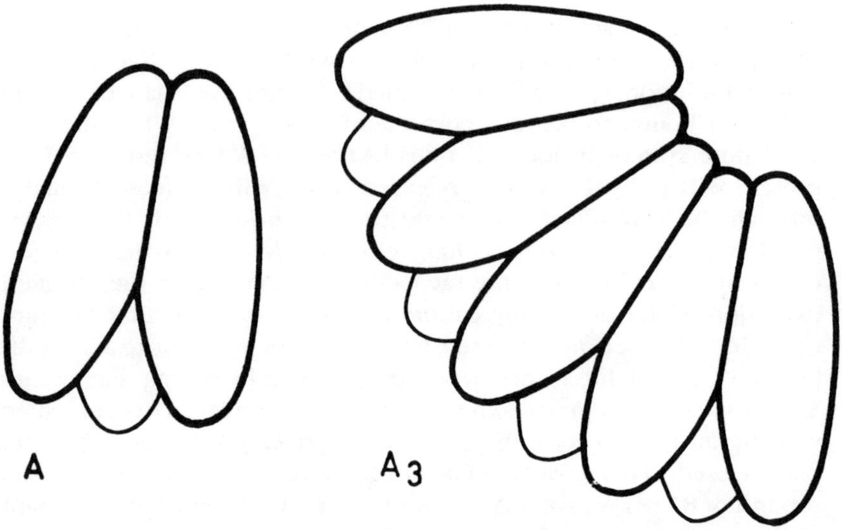

Fig. 8.5. Diagram illustrating early stage in the polymerization of TMV protein. A is A protein, a trimer of the ultimate protein subunit, and A_3 is a trimer of A.

monomer to form a protein tetramer. Critical analysis of osmotic pressure and sedimentation equilibrium measurements made by Shalaby et al. [9] showed that the isodesmic model as originally proposed by Banerjee and Lauffer fits the process better than any alternative models tested. As polymerization proceeds, a stable product with a sedimentation coefficient of approximately 20S is formed. Caspar [6] identified this with the double disc discovered by Markham et al. [10]. Double discs can stack to form rodlike structures, or, by another mode of polymerization, 20S material can polymerize further to form rods with a helical arrangement of the protein subunits. This fact led Lauffer [1] to postulate the existence of a two-turn spiral, called the double spiral, indistinguishable by sedimentation analysis or electronmicroscopy, from the double disc, but with a different type of bonding between subunits. These ideas about the polymerization process are discussed more fully by Lauffer [1, 11] and are illustrated in Figures 8.5 and 8.6.

Early studies, reviewed previously [1], confirmed the endothermic nature of the polymerization by direct calorimetry, even confirming the value for ΔH originally obtained by Banerjee and Lauffer [7] from equilibrium studies. It was shown that lowering pH promotes polymer-

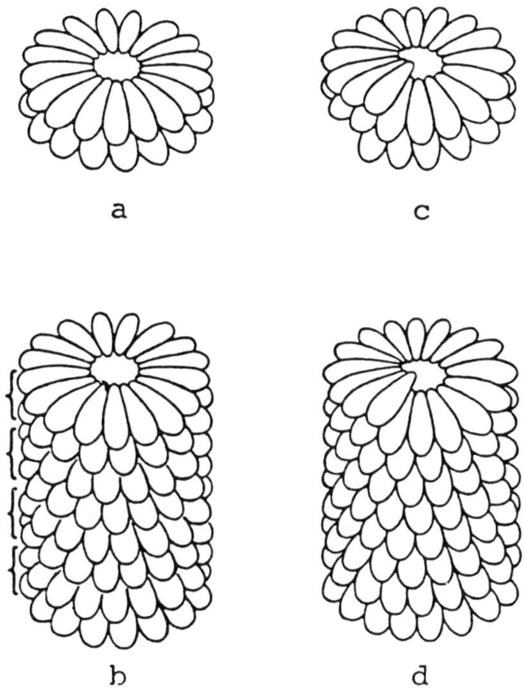

Fig. 8.6. Models of four aggregates of tobacco mosaic virus protein. **a:** The double disc. **b:** The stacked double disc (brackets identify individual double discs). **c:** The hypothetical double spiral. **d:** The helical rod. From Lauffer [1], reprinted with permission of the copyright owner, Springer Verlag.

ization; increasing ionic strength also promotes polymerization. Sucrose favors polymerization. Acetamide, thiourea, potassium thiocyanate, EDTA, urea, dioxane, and tetra-n-butylammonium bromide all favor depolymerization. Substituting D_2O for H_2O favors polymerization. Hydrogen ion titration measurements on both polymerized and unpolymerized TMV protein showed that hydrogen ions are bound during the polymerization process. Experiments carried out with ion-specific electrodes and with anion and cation exchange membrane electrodes demonstrated that TMV protein does not bind Cl^-, K^+ or Ca^{2+} ions during the polymerization process. All of these results were reviewed in greater detail previously [1] and original references were cited.

Since the publication of *Entropy-Driven Processes in Biology* [1], the author and his colleagues, especially Dr. Ragaa A. F. Shalaby, have given

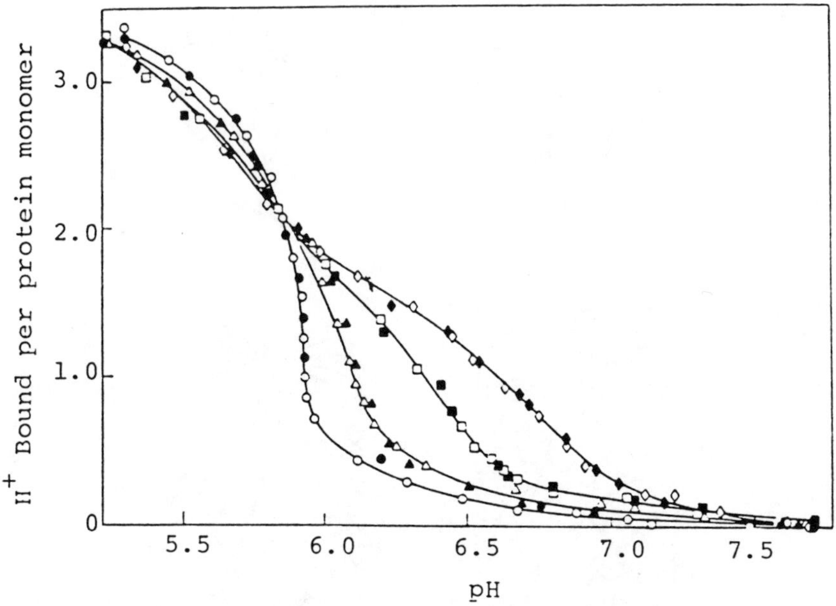

Fig. 8.7. Acid-base titration curves of TMV protein in 0.1 m KCl. Open symbols, HCl; closed symbols, KOH. ○, ●, 4°C; △, ▲, 10°C; □, ■ 15°C; ◇, ◆ 20°C. From Shalaby and Lauffer [12], reprinted with permission of the copyright owner, Academic Press, Inc.

careful attention to the early stages in the polymerization of TMV protein and of strains of the protein with known different amino acid compositions. They reinvestigated the hydrogen ion uptake upon polymerization of the protein, giving special attention to assuring that the process was completely reversible [12]. The results of hydrogen ion titration experiments at four temperatures between 4°C and 20°C in the pH range 5.3 to 7.7 are shown in Figure 8.7. In parallel experiments with the ultracentrifuge, the fraction of the protein remaining in the state sedimenting at approximately 4S was determined at the four temperatures and at four different pH values. At 4°C most of the material was in the 4S state at pH values between 6.05 and 6.42 and 100% at pH 6.56. At 20°C, 100% of the protein was in a polymerized state at pH 6.05 and 6.19. At pH 6.56 and 20°C, 31% of the protein remained in the essentially unpolymerized state, and the balance was distributed between two polymers, one sedimenting at 24S and one at 28S. By measuring the vertical distance in Figure 8.7 between the titration curve at 4°C, the lowest curve, and that at 20°C, the highest curve, it was established that at pH 6.56 1.08 more moles of hydrogen ion were bound per protein unit at 20°C than at 4°C. When the amount remaining unpolymerized was taken into consideration, this result shows that approximately 1.5 moles of hydrogen ion are bound per mole of protein during polymerization. This corresponds to an uptake of 4.5 hydrogen ions per A-protein (trimer) particle.

Shalaby and Lauffer [12] performed a second kind of experiment. An unbuffered solution at a concentration of 41 gm/l in 0.10 M KCl was shown by ultracentrifugation experiments to consist of 75% 20S material and 25% 8S material. The pH was 7.18. When this solution was diluted with 0.1 M KCl adjusted to pH 7.18, there was no change in pH, and all of the material was found in the 4S state. When the result of this experiment is contrasted with the result from the titration experiment, it is clear that there are two modes of polymerization of TMV protein. One at pH values between 6 and 7, involving the binding of hydrogen ions, and one at pH values above 7, involving no hydrogen ion binding. Shalaby and Lauffer interpreted this to mean that above pH 7, polymerization resulted in the formation of double discs, model a in Figure 8.6, and below pH 7 polymerization resulted in the formation of a different structure, assumed to be the double spiral or two-turn helix illustrated in model c of Figure 8.6.

During the course of experimentation on TMV protein polymerization at pH values below 7, the principal factors affecting the extent of polymerization were identified. They are A-protein concentration, temperature, hydrogen ion binding as a function of pH, ionic strength, and dipolar ion concentration. Analysis of the effect of ionic strength, to be described later, led to the conclusion that two considerations are involved; one is something akin to salting-out of proteins, and, because the polymerizing particles are charged, one involves an electrical work contribution to the free energy change. For experiments in which the polymerization is limited to the transition from the 4S state to the 20S state, the reaction can be considered to be

$$nA + n'H^+ = P$$

Because of the introduction of the electrical work term ΔW_{el}, equation 8.11 must be modified to assume the form of equation 8.14.

$$\ln K_a = \Delta S°/R - \Delta H°/RT - \Delta W_{el}/RT \qquad (8.14)$$

The mass action equation for the reaction can be written

$$K_a = [P]/\gamma^n[A]^n a_{H+}^{n'} \qquad (8.15)$$

In this equation [P] is the molar concentration of polymer, assumed to be a double spiral, [A] is the molar concentration of A protein (trimer), and a is activity. Since dipolar ions affect polymerization in a manner comparable with salting-out, it is assumed that the activity coefficient for [A], γ, is related to ionic strength and concentration of dipolar ion by equation 8.16.

$$\log \gamma = K_s'\mu + K_d m \qquad (8.16)$$

K_s' is the salting-out constant, K_d is a comparable constant for the effect of dipolar ions, μ is ionic strength, and m is the molar concentration of dipolar ion. Because P is not salted out, the activity coefficient for [P] in equation 8.15 is assumed to be unity. The validity of this assumption is strengthened by the fact that [P] is of the order of magnitude of 10^{-6} molar in most experiments. By substituting equations 8.14 and 8.16 and the definition of pH into equation 8.15 and by converting all logarithms to the base 10, one obtains equation 8.17.

$$\Delta S°/R - \Delta H°/RT - \Delta W_{el}/RT = 2.3 \log [P] - 2.3\, n \log [A] - 2.3\, n\, K_s'\, \mu - 2.3\, nK_d m + 2.3\, n'pH \qquad (8.17)$$

Two kinds of experiment, to be interpreted in terms of equation 8.17, were performed. In light-scatter experiments, turbidity or optical density, OD, which is proportional to turbidity, increases when temperature increases because average molecular weight is higher at higher temperatures. The results shown in Figure 8.8 illustrates this. These experiments have the advantage of being easy to perform, permitting the accumulation of much information, but they have the disadvantage of some thermodynamic ambiguity. Thermodynamic parameters like ΔS

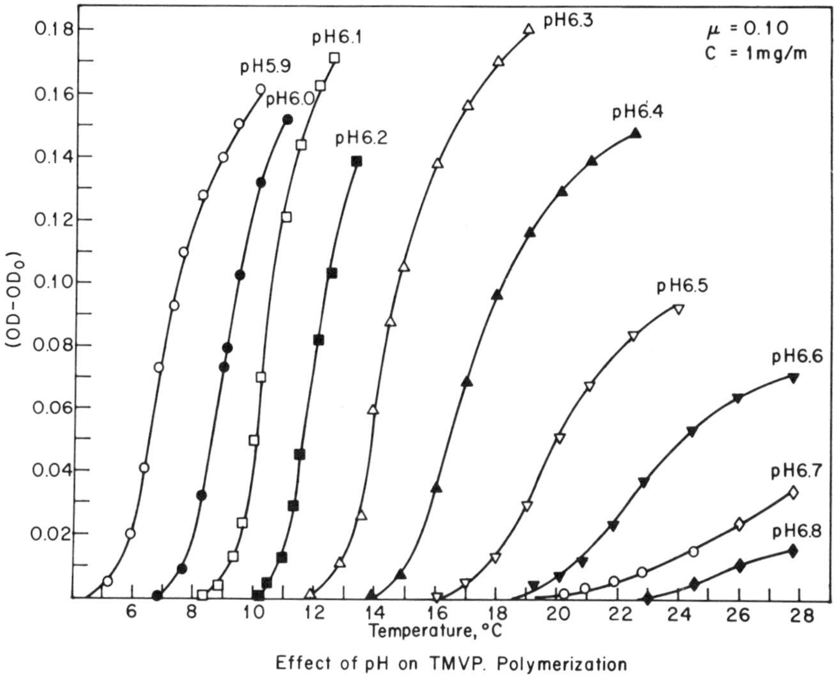

Fig. 8.8. The effect of pH on TMV protein polymerization indicated by $(A - A_o)$ plotted against temperature at $\mu = 0.10$. From Lauffer and Shalaby [13], reprinted with permission of the copyright owner, Academic Press, Inc.

and ΔH depend only on initial and final states in the reaction process. Strict application of equation 8.17 depends on all experiments having the same initial state and the same final state. Light-scatter experiments, as carried out in the author's laboratory, do not provide information about initial and final states; it is an assumption that they are the same in all experiments. When one solves equation 8.17 for $1/T^*$ after having made the definitions described below, one obtains equation 8.18

$$1/T^* = [1/(\Delta H^* + \Delta W^*_{el})][\Delta S^* - (R/n)\ln K^*_c + 2.3\ RK'_s\ \mu + 2.3\ RK_d m - 2.3\ R\xi\ pH] \qquad (8.18)$$

In equation 8.18, K_c is defined as $[P]/[A]^n$ and K^*_c is a constant value of K_c corresponding to a very low increment in OD, arbitrarily chosen to be 0.01. T^* is the temperature on the Kelvin scale at which that characteristic value of K_c^* is obtained. The thermodynamic parameters ΔS^*, ΔH^*, and ΔW_{el}^* are $\Delta S°$, and $\Delta H°$, ΔW_{el}^o each divided by n, in other words, the parameters corresponding to the consumption of one mole of A protein rather than those corresponding to the production of one mole of P. Finally, ξ is defined as n'/n. The derivation of equation 8.18 is described more fully by Lauffer and Shalaby [13].

A more rigorous experimental approach, introduced by Shalaby, et al. [9], involves determining the loading concentration of TMV protein at which the first trace of 20S component appears in the ultracentrifuge. This type of experiment has the advantage of being thermodynamically rigorous. The original and final states in all experiments are definitely the 4S and the 20S components and reversibility can be ascertained. However, the experiments have the disadvantage of being tedious. To obtain the loading concentration at which the 20S component first appears, it is necessary to carry out a series of ultracentrifuge experiments with different loading concentrations of A protein and pick the lowest concentration at which the 20S material is visible. When c_A/M_A is substituted for [A] and β/nM_A is substituted for [P], when c_A is set equal to the loading concentration in gm/l, c, minus the constant barely detectable concentration of P, β, and when the thermodynamic parameters have the same meaning as in equation 8.18, equation 8.17 can be transformed into equation 8.19.

$$\log(c - \beta) = \Delta H^*/2.3RT + \Delta W_{el}^*/2.3RT - \Delta S^*/2.3R + (1/n)\log\beta - (1/n)\log nM_A + \log M_A - K_s'\mu - K_d m + \xi pH \qquad (8.19)$$

It should be noticed that the third through sixth terms on the right of equation 8.19 are constants. In its general form, in which a constant is substituted for these terms, equation 8.19 was shown to be model-free. Except for the anatomy of the constant, the same equation can be derived from the thermodynamics of phase transition in which 4S TMV protein, A, is treated as a solute and the 20S component, P, as an insoluble phase in equilibrium with solute [9]. Another approach that leads to equation 8.19 in its general sense was derived for a model consisting of A protein in idodesmic equilibrium with A_2, A_3, etc. and simultaneously in equilibrium with the 20S component, P [9]. The only difference between these various derivations of equation 8.19 is in the anatomy of the constant term. Figure 8.9 illustrates how c is determined.

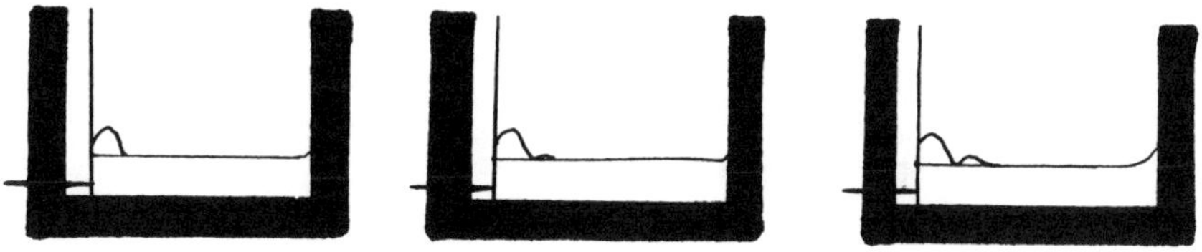

Fig. 8.9. Sedimentation schlieren patterns of TMV protein in 0.1 μ phosphate buffer, pH 6.7, at 15°C. Centrifugation was carried out at 40,000 rpm with a bar angle of 70°. Reprinted from Shalaby et al. [9], with permission of the copyright owner, Academic Press, Inc.

In the sedimentation diagram on the left, only the large amount of 4S material is apparent. The diagram on the right shows, in addition to the 4S component, a significant amount of 20S component. The one in the middle shows just a hint of the 20S component and therefore the value of c for that ultracentrifuge run is taken as the end point for substitution into equation 8.19. The value of the constant, β, the least detectable value of [P], was obtained by measuring c_P, the area under the 20S boundary, for various high loading concentrations, c, and then extrapolating to the loading concentration at which the 20s boundary was just visible. In this manner, β was found to be 0.33 gm/l.

Of all of the terms in equation 8.19, the most difficult one to deal with conceptually is the electrical work term, the second on the right. That there should be an electrical work term is obvious because TMV protein is charged both in the 4S and the 20S states at the pH values used in polymerization experiments. Determination of the proper theoretical approach for understanding electrical work terms is not easy, because the theories are model-dependent. A general feature of such theories is that the electrical work term depends upon the square of the electrical charge and is in some way inversely related to ionic strength. When an experiment is carried out at constant temperature and constant pH in the absence of dipolar ions, but at varying values of ionic strength, μ, equation 8.19 reduces to equation 8.20.

$$\log(c - \beta) = \text{const.} + \Delta W^{*}_{el}/2.3\ RT - K'_s\ \mu \qquad (8.20)$$

The results of such an experiment are shown in Figure 8.10. The fact that the graph of log (c − β) is not a straight line when plotted against μ is evidence that some term other than salting-out is significant. That term is by the present theory $\Delta W^{*}_{el}/2.3$ RT. The data in this experiment

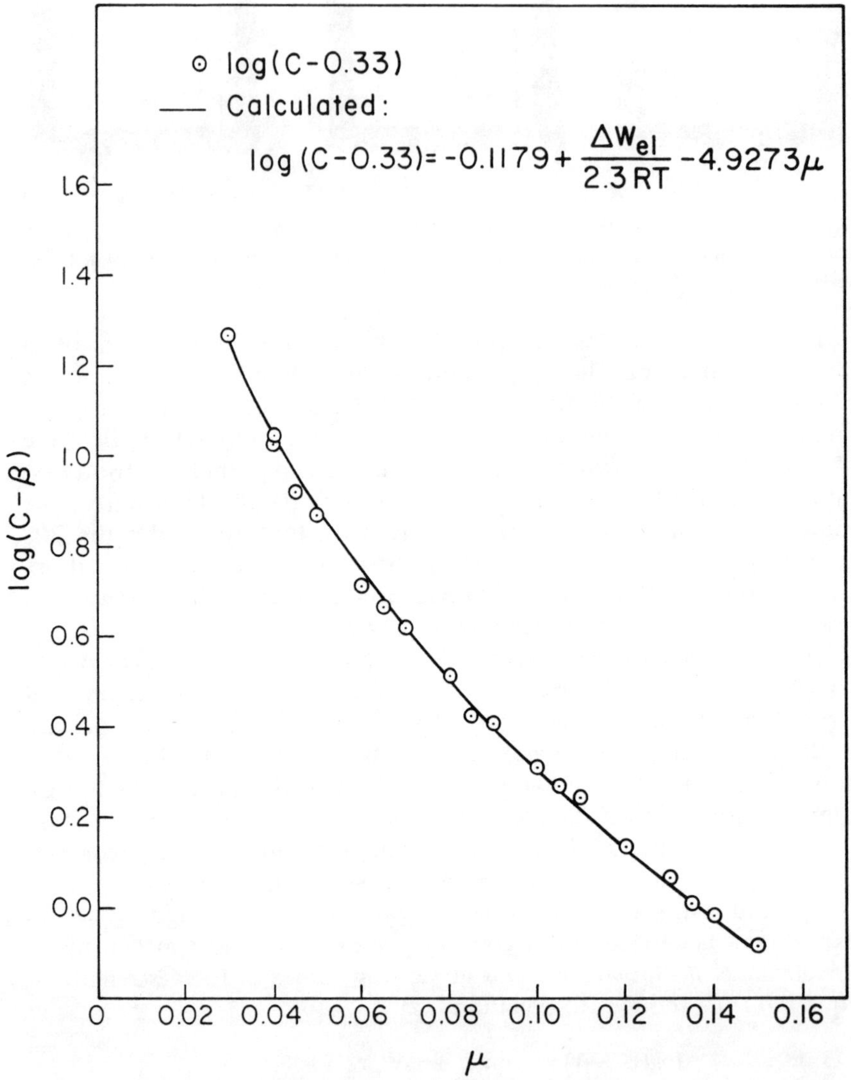

Fig. 8.10. Log (c − 0.33) that gives the smallest 20S peak as a function of ionic strength at pH 6.7 and 15°C. Reprinted from Shalaby et al. [9], with permission of the copyright owner, Academic Press, Inc.

were fitted to the equation, $\log(c - 0.33) = -0.1179 + 0.326/(\mu^{1/2} \exp[1/29.9\, \mu^{1/2}]) - 4.93\, \mu$. The curved line fitting the data of equation 8.20 is a graph of this semiempirical equation. The salting-out constant, K_s', is identified as 4.93. In the electrical work term, $\Delta W_{el}^*/2.3\ RT$, is identified with the term involving $\mu^{1/2}$. At 15°C, pH 6.7, and ionic strength 0.10, ΔW_{el}^* has the value of 1.22 kcal/mole of A protein.

Confirmatory evidence for the existence of an electrical work term was obtained from light-scatter experiments. Inspection of equation 8.18 shows that experiments carried out under conditions in which everything is held constant except ionic strength, μ, should give a straight line when $1/T^*$ is plotted against μ if ΔW_{el}^* is independent of μ. However, the results of such experiments on TMV protein yielded curved relationships between $1/T^*$ and μ that could be interpreted in terms of a value of ΔW_{el}^* that varied with μ. Similar experiments were performed on the proteins of the E66 [16] and the flavum strains of TMV [17]. The proteins of these two strains have amino acid compositions very similar to that of TMV protein but each with a net charge at pH 7.5 of −3 per protein monomer compared with −4 for TMV protein. Since, as is explained in a footnote,[1] the electrical work effect is proportional to the square of the charge, the electrical work terms for these two proteins calculated with the aid of equation 8.18 turned out to be nearly equal, and both were approximately half that calculated the same way for TMV protein. Furthermore, the amino acid compositions show that E66 protein is less hydrophobic than TMV protein, and flavum protein is more hydrophobic than TMV protein. Since salting-out constants are correlated with hydrophobic nature, one should expect a lower salting-out constant for E66 protein and a higher salting-out constant for flavum protein than for TMV protein. The analyses of the data yielded values consistent with this understanding. However, the actual values of the salting-out constant K_s' and the electrical work term ΔW_{el}^* for TMV protein were somewhat different from those obtained by analysis of the ultracentrifuge data.

Shalaby and Lauffer [18] reinvestigated the effect of ionic strength on the polymerization of E66 protein by the ultracentrifuge method. Results similar to those displayed in Figure 8.10 were obtained. When analyzed in accordance with equation 8.19, the data yielded a value of ΔW_{el}^* at pH 6.7, 15°C, and $\mu = 0.10$ of 0.700 kcal/mole for E66 protein, compared with 1.22 for TMV protein. The experimental ratio was 0.574, to be compared with $(-3/-4)^2$, or 0.5625. K_s' was calculated to be 2.16 for E66, compared with 4.93 for TMV protein, consistent with the less hydrophobic nature of the E66 protein. However, similar experiments carried out on the protein of the ribgrass strain of TMV failed to provide evidence of an electrical work term [19]. Since this protein is more highly charged than the TMV protein, one would expect a higher electrical work term than for TMV protein. Taken at face value, this result contradicts the earlier experience. However, only 60 mg of ribgrass virus protein were available for the experiments. This made it necessary to restrict ultracentrifugation runs to those with low loading concentrations, values of $\log(c - \beta)$ of 0.59 and lower. A straight line could be fitted to the data of Figure 8.10 for values of $\log(c - \beta)$ of 0.4 and lower. The convincing evidence for the existence of an electrical work term with TMV protein comes from inclusion of data at higher loading concentrations. Thus the negative result with the ribgrass virus protein could stem from the inability to perform experiments at higher concentrations. Overall, the experience indicates that the analysis of the effect of ionic strength involved in the development of equations 8.18 and 8.19 is, at least in principle, correct.

Shalaby et al. [9] determined the effect of temperature on the polymerization of TMV protein by the ultracentrifugation method. The results are shown in Figure 8.11. Because one can work only over a limited temperature range at any given pH value, it was necessary to carry out experiments at three different values of pH and then combine the data by simply adjusting the vertical coordinates by a constant amount for each pH value. Since, as was pointed out earlier (see footnote 1), $\Delta W^*_{el}/T$ is practically invariant with T, when μ, M, and pH are kept constant in an experiment, $\log(c - \beta)$ plotted against 1/T should give a straight line with a slope of $\Delta H^*/2.3\ R$. Except for two data obtained at the highest temperatures, the results displayed in Figure 8.11 fit a straight line admirably. From the slope, ΔH^* was evaluated to be +32 kcal/mole of A protein. Similar experiments on E66 protein [18] yielded values for ΔH^* per mole of A protein of 33 kcal at pH 6.7 and 36 kcal at pH 6.9. It is interesting to observe that the constant value of ΔH^* fits the data obtained with TMV protein over a temperature range of 20°C. Such a result can be obtained only when the heat capacity of the 4S component is the same or very nearly the same as that of the 20S component. For many other entropy-driven processes reported in the literature, ΔH^* varies considerably with temperature.

The experiments of Banerjee and Lauffer [7] carried out with the osmometer in which the polymerization was limited to the formation of very small polymers, A_2 and A_3, when interpreted in terms of the

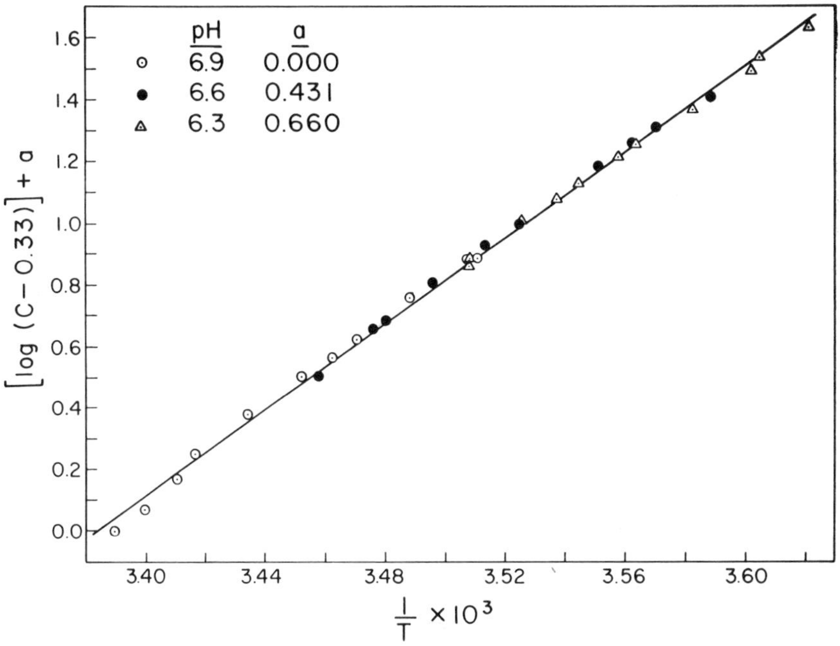

Fig. 8.11. Log (c − 0.33) + α that gives the smallest visible 20S peak as a function of reciprocal of temperature in Kelvin at pH 6.3, 6.6, and 6.9 at ionic strength of 0.1; α is a constant to shift the ordinate of the data at the different pH values. Reprinted from Shalaby et al. [9], with permission of the copyright owner, Academic Press, Inc.

mathematics of an isodesmic mechanism, yielded a value of 30 kcal per mole. Direct calorimetric measurements by Stauffer, et al. [20] for the dissociation of the 20S component to the 4S component at pH 7.5 yielded confirming values ranging between −25 and −30 kcal/mole of A protein. Heat was released in these dilution experiments; this corresponds to positive values of the enthalpy for polymerization.

The results obtained in ultracentrifugation experiments when pH is the only variable are shown in Figure 8.12 [9]. The slope of this line is 1.22. However, this cannot be taken to be a true value of the ξ appearing in equation 8.19, because the electrical work term, $\Delta W^*_{el}/2.3\ RT$, varies with charge and therefore with pH. A correction must be made to take this into account. The details of this correction are discussed by Shalaby, et al. [9]. The gist of the idea is that at pH values above 7, unpolymerized TMV A protein has a charge of −4 protonic units per protein subunit. As

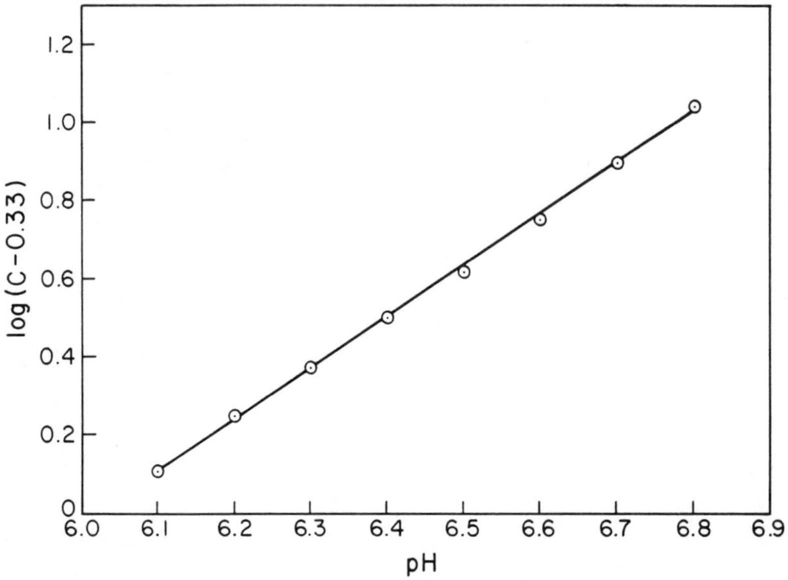

Fig. 8.12. Log (c − 0.33) at which the smallest visible 20S peak appears as a function of pH. The protein is in 0.1 μ phosphate buffer, and the temperature is 9.5°C. Reprinted from Shalaby et al. [9], with permission of the copyright owner, Academic Press, Inc.

Figure 8.7 shows, when the pH is reduced, the charge on the unpolymerized protein decreases. The titration curve at 4°C shows this. It is assumed that A_1 first binds ξH^+ and then polymerizes. Since the charge on A_1 is 3q initially, at the instant of polymerization it is $3q + \xi$ or $q + \xi/3$ per protein subunit. In this manner a correction in the electrical work term at different pH values was obtained and the true value of ξ of equation 8.19 was estimated to be 1.1. Inspection of the titration curve at 4°C shown in Figure 8.7 shows that at pH 6.5, the center of the range corresponding to the data of Figure 8.12, A protein still in the unpolymerized state has bound 0.14 hydrogen ions per protein subunit, or 0.42 per A protein unit. When this is added to the value of ξ, at the instant of polymerization the A unit has bound 1.52 hydrogen units, corresponding to 0.5 per protein subunit. Thus, at pH values below 7, polymerization is preceded by the binding of one hydrogen ion for every two protein subunits, to permit the formation of the 20S state, presumably the two-turn helix illustrated as c in Figure 8.6. In contrast to this result, the titration curves shown in Figure 8.7 indicate that 1.5

hydrogen ions are bound per TMV protein subunit or 4.5 per A unit during polymerization. Analyses of the centrifugation patterns carried out by Shalaby and Lauffer [12] show that at pH 6.42, for example, approximately two-thirds of the protein sediments at a rate of 129 svedbergs. Thus the resolution of the fact that only 0.5 hydrogen ions must be bound as a prerequisite for polymerization to the 20S state but that substantially more hydrogen ions are bound when the temperature is raised from 4°C to 20°C is that additional hydrogen ions are bound in polymerization beyond the 20S state. The data of Figure 8.7 establish that, at pH values near 6.0, where ultracentrifugation analyses [12] show that all of the protein is in a highly polymerized state at 20°C, two hydrogen ions have been bound per protein subunit.

There are no groups on the TMV protein that normally bind hydrogen ions in the neighborhood of pH 6.5. Fraenkel-Conrat and Ramachandran attributed this binding to abnormal carboxyl groups [21] and Caspar [22] identified these abnormal carboxyl groups with lead binding sites located 25 Å and 84 Å from the axis of the TMV particle.

Shalaby and Lauffer [23] speculated on the identity of the abnormal carboxyl group responsible for binding hydrogen ions in the polymerization to the 20S state in light of the X-ray diffraction studies of Bloomer et al. [24] and of Stubbs et al. [25] and concluded that glutamic acid 145 is a likely prospect. Butler and Durham [26] had suggested this previously. The 25-Å site could be in the carboxyl cage described by Stubbs et al. [25] for the virus.

Experiments just discussed show that when TMV protein is polymerized in the presence of a reservoir of hydrogen ions such as that supplied by a buffer at pH values below 7, one hydrogen ion is bound for every two protein subunits in going from the 4S to the 20S state. However, as was already mentioned, when 20S material at a pH value above 7 at the high concentration is depolymerized to 4S material by dilution, no hydrogen ions are released, indicating that there is a form of 20S material that does not bind hydrogen ions upon polymerization. Further experiments were carried out by Shalaby and Lauffer [23]. TMV protein was polymerized at pH values above 7 in unbuffered solution, either by increasing ionic strength at constant temperature or by raising temperature at constant ionic strength. In both cases, a 20S component is formed that has bound only a very small fraction of the amount of H^+ ion bound when a reservoir of such ions is available. This small amount comes from the slight buffering capacity of the A protein itself. Thus there are two forms of 20S protein, one of which has bound one hydrogen ion per two protein subunits and one of which has not. When the 20S material that has not bound hydrogen ions is titrated with HCl, it yields a 20S form that has bound one hydrogen ion per two subunits. Shalaby and Lauffer [23] identify the 20S form that has not bound hydrogen ions with the double disc illustrated in a of Figure 8.6 and the form that has bound the hydrogen ion per two subunits with the structure illustrated in c of that figure.

When all of the facts known about hydrogen ion binding by TMV protein are put together, a somewhat complex situation emerges. Protein subunits polymerize to form A protein, trimers of the subunit, without binding hydrogen ion. When these polymerize in the absence of a reservoir of hydrogen ions at pH values above 7, a 20S form, the double disc, is obtained without binding hydrogen ions. When hydrogen ions are available in a reservoir or made available by titration, even at pH values above 7, one hydrogen ion is bound for every two protein subunits in the 20S form, presumably the double spiral. When pH is lowered by titration, still more hydrogen ions are bound, and further polymerization takes place. When pH 5.9 is reached, the protein is fully polymerized in the helical state with two hydrogen ions bound per protein subunit. Investigations carried out in a number of laboratories are consistent with this view [27, 28].

England and Cohn [29] pointed out that dipolar ions behave like salting-out agents. Lauffer and Shalaby [30] showed that bovine serum albumin promotes the polymerization of TMV protein. These facts led to the inclusion of the K_dM term in equations 8.18 and 8.19. A more thorough investigation of the effect of dipolar ions, glycine, glycylglycylglycine, and glycylglycylglycylglycine, was carried out by Lauffer and Shalaby [31]. The effect of glycine as investigated by the light-scatter method is shown in Figure 8.13, and the effect of glycylglycine as investigated by the ultracentrifuge method is shown in Figure 8.14. When $\log(c - \beta)$ is plotted against the molarity of glycylglycine, with the exception of two points on the graph, the data fall on a straight line, consistent with theory expressed by equation 8.19.

Considerable literature, reviewed by Gallagher and Lauffer [32], deals with the question of the binding of cations, particularly divalent cations, to TMV. There are two such binding sites per protein subunit, already referred to as lead binding sites. When cations, particularly divalent cations, are bound at these sites, hydrogen ions are released. The same sites are regarded as being responsible for the binding of hydrogen ions in the pH range 6–7. The sites are regarded as being carboxyl groups,

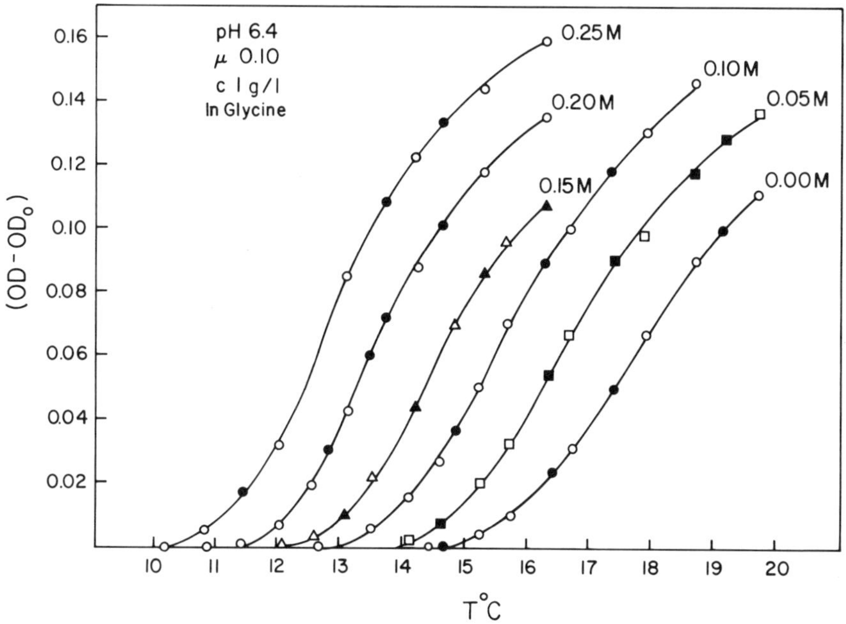

Fig. 8.13. T vs. ($OD - OD_o$) at 320 nm for the polymerization of TMV protein at pH 6.4, ionic strength 0.10, and protein concentration of 1 gm/l at different molar concentrations of glycine. Open symbols represent increasing temperature, and closed symbols represent decreasing temperature. Reprinted from Lauffer and Shalaby [31], with permission of the copyright owner, Academic Press, Inc.

which, because of their environment, have their pK values shifted upward. Gallagher and Lauffer [33] found, in contrast, that polymerized TMV protein does not bind calcium ions under equilibrium conditions at pH values above the isoionic point, pH 4.3–4.6. However, calcium ions are bound under nonequilibrium conditions. This nonequilibrium binding is observed when a solution of helically polymerized coat protein is titrated with $CaCl_2$ immediately after raising pH from the isoionic point to higher values. At pH 5.6 and 4°C, equilibrium values, indicating no calcium ion binding, were established in about 3 hours. Thus it can be concluded that at the isoionic point, polymerized TMV protein does have a structure capable of binding calcium ions, but this structure gradually relaxes at higher pH values into one incapable of such binding. This result is somewhat comparable with an earlier finding of McMichael and Lauffer [34]. When TMV protein stored in the

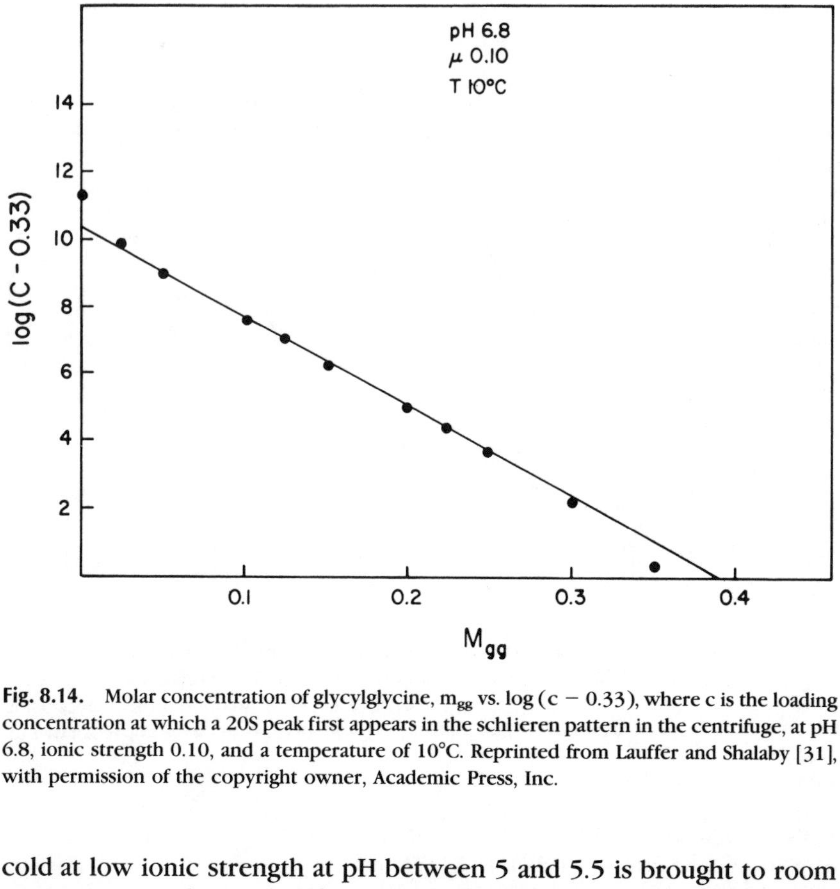

Fig. 8.14. Molar concentration of glycylglycine, m_{gg} vs. log (c − 0.33), where c is the loading concentration at which a 20S peak first appears in the schlieren pattern in the centrifuge, at pH 6.8, ionic strength 0.10, and a temperature of 10°C. Reprinted from Lauffer and Shalaby [31], with permission of the copyright owner, Academic Press, Inc.

cold at low ionic strength at pH between 5 and 5.5 is brought to room temperature and mixed with acetate buffer and additional electrolyte to give a final pH of 6.5 and ionic strength of 0.1, the protein is still polymerized. When the temperature is then dropped to 5°C, the protein depolymerized rapidly unless the added electrolyte contains 0.01 m$CaCl_2$. Under these circumstances, depolymerization is slow, requiring about 12 hours to reach completion, but when the thus depolymerized protein is repolymerized by raising the temperature, this repolymerized protein depolymerizes rapidly when temperature is again lowered, even though calcium ions are present. Ion binding experiments showed, in this case, too, that calcium ion is bound during the initial step when the pH is brought to pH 6.5 at room temperature in the presence of that ion. The calcium ion is released during the slow depolymerization when the temperature is dropped and is not bound again during repolymerization at pH 6.5. This result also suggests a structure of polymerized protein at low pH values capable of binding divalent cations, a structure that relaxes at higher pH values.

SIGNIFICANCE OF ENTROPY-DRIVEN PROCESSES IN BIOLOGY

The X-ray diffraction analyses of Bloomer et al. [24] on TMV double discs and of Stubbs et al. [25] on the virus itself establish clearly that ordinary weak bonds are involved in stabilizing the polymeric structures. These bonds are of the same nature as those that stabilize structures produced by enthalpy-driven processes. The big difference between entropy-driven processes and enthalpy-driven processes is that in the former, water molecules are released as part of the reaction; the polymer is less highly hydrated than the polymerizing units. Reflection on the nature of the entropy-driven assembly processes in biological systems, to be discussed in the next chapter, leads to a conclusion that entropy-driven processes occur in dynamic systems, systems in which structures must be formed from subunits when needed and disassembled when their function has been served. In the case of tobacco mosaic virus, the coat protein must be readily removed from the RNA during the infection process.

A major question is why entropy-driven processes were selected in evolution for dynamic assembly-disassembly reactions in biology. An answer is provided in an essay written by the author [35]. The total free energy change for the polymerization of TMV A protein is the sum of several terms, as indicated by equation 8.21.

$$\Delta G = \Delta G_b + \Delta G_w + \Delta W_{el} + \Delta G_{H+} + \Delta G_o \qquad (8.21)$$

In this equation, $_b$ indicates bond formation, $_w$ indicates water release during assembly, $_{H+}$ means the binding of hydrogen ions, and $_o$ means all other known and unknown factors such as the role played by various chemicals, etc. ΔG_b is a large negative contribution to the net free energy, and ΔG_w is a large positive contribution. Their sum is a small number. ΔW_{el}, ΔG_{H+}, and ΔG_o are relatively small, but they are important for regulating the reaction. When polymerizing units and polymers are in equilibrium, net ΔG is zero. Small changes in these small contributions can make ΔG negative, in which case assembly proceeds, or positive, in which case structures come apart. However, since the first two terms on the right are large, for the sake of simplicity in further discussion attention will be focused on them alone.

Almost all biological processes take place at temperatures of 300 K± 10%. At temperatures below that range, most organisms freeze, and at temperatures above that range biological systems become inactivated, except in a few thermophilic organisms. For systems at equilibrium at 300 K, when $\Delta G = 0$, equation 8.22 can be written

$$\Delta H = T\Delta S \simeq 300\Delta S \tag{8.22}$$

Since, for purposes of discussion, attention is focused on bond formation and water release, as approximations, equations 8.23 and 8.24 can be written

$$\Delta H = \Delta H_b + \Delta H_w \tag{8.23}$$

$$\Delta S = \Delta S_b + \Delta S_w \tag{8.24}$$

It is general experience that when bonds are formed between free units to join them together, the change in entropy resulting from loss of vibrational and rotational freedom is of the order of magnitude of −100 cal/mole deg. Theoretical calculations show that this order of magnitude is not highly sensitive to the size of the assembling units. Be that as it may, one can write equation 8.25 where $-s$ is simply the value of the entropy decrease for bonding between two polymerizing bodies.

$$\Delta S_b = -s \tag{8.25}$$

The entropy change for water release, ΔS_w, is positive. It can be less than or greater than s in magnitude. This can be indicated symbolically by equation 8.26.

$$\Delta S_w = +xs \tag{8.26}$$

In this equation, x is a variable equal to or greater than 0, a variable proportional to how much water is released in the reaction. Equation 8.27 results when the last three are combined.

$$\Delta S = -s + xs = -s(1 - x) \qquad (8.27)$$

Bull [36] carried out extensive studies on the equilibrium between several proteins and water vapor at different activities and at different temperatures. From these results he was able to estimate ΔH and ΔS for water sorption and conversely for water release. For these various proteins, ΔH was many hundred times ΔS. This idea can be represented symbolically by equation 8.28, into which equation 8.26 can be substituted to yield the final term on the right.

$$\Delta H_w \simeq 500\Delta S_w = 500xs \qquad (8.28)$$

The accuracy of the number 500 is not important to the logical development that follows, as long as the actual number is substantially greater than 300, biological temperature. When the value for ΔS given by equation 8.27 is substituted into equation 8.22, equation 8.29 results.

$$\Delta H = -300s(1 - x) \qquad (8.29)$$

When equation 8.23 is solved for ΔH_b and values for ΔH_w and ΔH from equations 8.28 and 8.29 are substituted, equation 8.30 results.

$$\Delta H_b = \Delta H - \Delta H_w = -300s(1 - x) - 500xs = -s(300 + 200x) \qquad (8.30)$$

From equations 8.25 through 8.30, the hypothetical thermodynamic parameters, shown in Table 8.1 for systems that can be reversed at 300 K, can be calculated for various values of x, the function that is proportional to the amount of water released in the assembly process. For values of x between 0 and 1, ΔH_b, the parameter that measures the strength of the bond between polymerizing units, varies between −300 *s* and −500 *s*. The total ΔH is negative; the reaction is exodermic or enthalpic. When x has a value of 1, ΔH_b has a value of −500 *s* and ΔH is 0; the process is temperature independent. For values of x greater than 1, ΔH_b increases in magnitude from −500 *s* to −1,100 *s* when x is

TABLE 8.1. Hypothetical Thermodynamic Parameters for Systems in Equilibrium at 300 K

x	ΔH_b	ΔH_w	ΔH	ΔS_b	ΔS_w	ΔS	$-300\Delta S$	ΔG	Reaction type
0	$-300s$	0	$-300s$	$-s$	0	$-s$	$+300s$	0	Enthalpic
1	$-500s$	$+500s$	0	$-s$	$+s$	0	0	0	Temperature independent
2	$-700s$	$+1{,}000s$	$+300s$	$-s$	$+2s$	$+s$	$-300s$	0	Entropic
3	$-900s$	$+1{,}500s$	$+600s$	$-s$	$+3s$	$+2s$	$-600s$	0	Entropic
4	$-1{,}100s$	$+2{,}000s$	$+900s$	$-s$	$+4s$	$+3s$	$-900s$	0	Entropic

4, signifying stronger and stronger bonding between polymerizing units. Total ΔH is positive; the reaction is entropic. Thus, the greater the water release upon association, the stronger can the bonds be between polymerizing units without sacrificing the property of ready reversibility at biological temperature. This makes possible the formation of precisely ordered structures with high tensile strength. Thus the advantage conferred by entropy-driven assembly processes in dynamic biological systems seems to be that such processes permit the formation of easily reversed but strong, precisely fitting structures.

It is necessary in view of these considerations to reassess the role of water in entropy-driven systems. It was formerly thought that the tendency for water molecules to dissociate from the polymerizing units at higher temperatures, akin to the melting of ice, was the driving force for polymerization [1]. In light of present knowledge it must be concluded that the driving force for the forward reaction is the formation of bonds, with the attendant negative contribution to the enthalpy (ΔH_b). The role of water is to facilitate the reverse process. A large release of water on association means a large binding of water on dissociation, something akin to what happens when proteins are dissolved.

REFERENCES

1. Lauffer, M.A. Entropy-Driven Processes in Biology. Berlin: Springer-Verlag, 1975.
2. Stevens, C.L. and Lauffer, M.A. Biochemistry *4:*31–37, 1965.
3. Lauffer, M.A., Ansevin, A.T., Cartwright, T.E., Brinton, C.C., Jr. Nature *181:*1338–1339, 1958.
4. Jaenicke, R. and Lauffer, M.A. Biochemistry *8:*3083–3092, 1969.
5. Ansevin, A.T. and Lauffer, M.A. Nature *183:*1601–1602, 1959.
6. Caspar, D.L.D. Adv. Protein Chem. *18:*37–121, 1963.
7. Banerjee, K. and Lauffer, M.A. Biochemistry *5:*1957–1963, 1966.
8. Durham, A.C.H. and Klug, A. J. Mol. Biol. *67:*315–332, 1972.
9. Shalaby, R.A., Stevens, C.L., and Lauffer, M.A. Arch. Biochem. Biophys. *218:*384–401, 1982.
10. Markham, R., Frey, S., and Hills, G.J. Virology *20:*88–102, 1963.
11. Lauffer, M.A. Biochemistry *5:*2440–2446, 1966.
12. Shalaby, R.A. and Lauffer, M.A. J. Mol. Biol. *116:*709–725, 1977.
13. Lauffer, M.A. and Shalaby, R.A. Arch. Biochem. Biophys. *201:*224–234, 1980.
14. Lauffer, M.A. In Subunits in Biological Systems, Part A, S.N. Timasheff and G.E. Fasman, (eds.). New York: Dekker, pp. 149–199, 1971.
15. Lauffer, M.A. Dev. Biochem. *3:*115–180, 1978.
16. Shalaby, R.A. and Lauffer, M.A. Arch. Biochem. Biophys. *204:*494–502, 1980.
17. Shalaby, R.A. and Lauffer, M.A. Arch. Biochem. Biophys. *204:*503–510, 1980.

18. Shalaby, R.A. and Lauffer, M.A. Arch. Biochem. Biophys. *236:*390–398, 1985.
19. Shalaby, R.A. and Lauffer, M.A. Arch. Biochem. Biophys. *238:*69–74, 1985.
20. Stauffer, J., Srinivasan, S., and Lauffer, M.A. Biochemistry *9:*193–200, 1970.
21. Fraenkel-Conrat, H. and Ramachandran, A.B.V. Adv. Protein Chem. *14:*175–229, 1959.
22. Caspar, D.L.D. Nature *177:*928, 1956.
23. Shalaby, R.A. and Lauffer, M.A. Arch. Biochem. Biophys. *223:*224–234, 1983.
24. Bloomer, A.C., Champness, J.N., Bricogne, G., Staden, R. and Klug, A. Nature *276:*362–368, 1978.
25. Stubbs, G.J., Warren, S.G., and Holmes, K.C. Nature *267:*216–221, 1976.
26. Butler, T.J.G. and Durham, A.C.H. J. Mol. Biol. *72:*19–24, 1972.
27. Schuster, T.M., Scheele, R.B., Adams, M.L., Shire, S.G., Steckert, J.J., and Totschka, M. Biophys. J. *32:*313–329, 1980.
28. Durham, A.C.H. and Finch, J.T. J. Mol. Biol. *67:*307–314, 1972.
29. England, A., Jr., and Cohn, E.J. J. Am. Chem. Soc. *57:*634–637, 1935.
30. Lauffer, M.A. and Shalaby, R.A. Arch. Biochem. Biophys. *178:*425–434, 1977.
31. Lauffer, M.A. and Shalaby, R.A. Arch. Biochem. Biophys. *242:*478–487, 1985.
32. Gallagher, W.H. and Lauffer, M.A. J. Mol. Biol *170:*905–919, 1983.
33. Gallagher, W.H. and Lauffer, M.A. J. Mol. Biol. *170:*921–929, 1983.
34. McMichael, J.C. and Lauffer, M.A. Arch. Biochem. Biophys. *169:*209–216, 1975.
35. Lauffer, M.A. Comments Mol. Cell Biophys. *2:*99–109, 1983.
36. Bull, H.B. J. Am. Chem. Soc. *66:*1499–1507, 1944.

NOTES

1. The electrical work theory used in this computation is based on Lauffer's theory for the electrical work of charging a cylinder with uniform immobilized charge on the cylindrical surface. The first publication [14] of this theory contained typesetting errors that were corrected in a later publication [15]. Like other theories for the charging of a cylinder, the electrical work is proportional to the square of the charge. Because of the assumption of fixed charges on the surface, there are two useful consequences. One is that the theory should apply to very short cylinders, like double discs or two-turn helixes. The other consequence is that the potential at the edge of the surface is less than that in the main part of the surface. This fact gives rise to an increase in electrical work when many small curved surfaces are assembled to form a larger one. This theory was developed into the form used above in Appendix B to the paper by Shalaby et al. [9]. A useful consequence of the theory is that $\Delta W^{*}_{el}/T$ is practically invariant with T.

9

Entropy-Driven Processes in Biology. II. Biological Applications

In Chapter 8, the entropy-driven polymerization of tobacco mosaic virus protein was examined in detail and the question was raised as to why in the process of evolution entropy-driven processes were selected for the functioning of organisms, especially for dynamic situations. The author's answer to this question, as developed in Chapter 8, is that entropy-driven structure-building processes permit easy reversibility at biological temperatures without sacrificing strength and architectural precision in the structures formed. The role of water is to facilitate disassembly of structures. Water is bound during disassembly; it must therefore of necessity be released during assembly. The increase in entropy in the assembly process results from this release of water.

In the present chapter, a large number of entropy-driven processes in biology are listed. Then a few selected dynamic processes are discussed. The processes selected are in reality highly complex in the living organism. The author has attempted to focus attention on the most important elements of each reaction, omitting most of the details. This is done solely for the purpose of highlighting the role of entropy-driven reactions. Much more nearly complete treatments are found in modern textbooks of biochemistry and molecular biology. Especially interesting and detailed accounts, in the view of the author, are found in the 1983 book, *Molecular Biology of the Cell,* by Bruce Alberts et al. [1].

Table 9.1 lists approximately 70 entropy-driven processes in biological systems. In every case the process listed has been shown to be entropy-driven by studies of the equilibrium constant at various temperatures. In most instances the Van't Hoff enthalpy was found to be positive. In a few, the enthalpies were negative but of low magnitude,

TABLE 9.1. Entropy-Driven Processes in Biology

Process	Reference
Polymerization or association of viral coat protein	
Tobacco mosaic virus (TMV) protein	2
Dahlemense strain TMV protein	2
Cucumber virus 4 protein	2
PM_2 mutant of TMV protein	2
E66 mutant of TMV protein	3,4
Flavum strain of TMV protein	3
Reconstitution of TMV	2
Tobacco rattle virus protein	2
Papaya mosaic virus protein	3
Alfalfa mosaic virus protein	3
Aggregation or crystallization of viruses	
Tobacco mosaic virus (TMV)	3
Carnation ringspot virus	3
Activities of cold-inactivated enzymes	
Urease	3
Glutamic acid dehydrogenase	3
Adenosine triphosphatase (ATPase)	3
Carbamyl phosphate synthetase	3
Glucose-6-phosphate dehydrogenase	3
D-β-hydroxybutyric acid dehydrogenase	3
Glycogen phosphorylase	3
A nitrogen fixing enzyme	3
Acetyl co-A carboxylase	3
Argino-succinase	3
17-β Hydroxysteroid dehydrogenase	3
Pyruvate carboxylase	3
Glutamic acid decarboxylase	3
Enzyme-substrate interactions	
Streptomycin trehalose inhibitor SGI with *Rhizopus* glucoamylase	5
D-glucose with yeast hexokinase	4
Bovine basic pancreatic trypsin (Kunitz) inhibitor with plasmin	6
Kazal inhibitor with β trypsin	7
Kazal inhibitor with trypsinage	7
Lactate dehydrogenase with four inhibiting dyes	8
Malate dehydrogenase with four inhibiting dyes	8
Substrate analogs to chemically modified α chymotrypsin	3
Oligamer formation in proteins and enzymes	
Polymerization of sickle cell hemoglobin	2
Carboxy hemoglobin A_o dimers association to tetramers (pH 8.6)	3
Dimerization of oxygenated α SH human hemoglobin chains	3
Tetramerization of oxygenated β SH human hemoglobin chains	3
Formation of hemoglobin from dimerization of oxygenated α-β chains	3

(Continued)

TABLE 9.1. Entropy-Driven Processes in Biology (Continued)

Process	Reference
Oligamer formation in proteins and enzymes (continued)	
Hexamer formation from physocyanin trimers	2
Dimerization of α chymotrypsin (low temperature)	3
Dimerization of bovine liver glutamate dehydrogenase (low temperature)	3
Complex formation between $\alpha_s 1$ and K-casein polymers	3
Association of reduced A and B chains to form toxic ricin	9
Tetramer formation in rabbit muscle phosphofructokinase (15–23°)	10
Association betwen ornithine carbamoyltransferase and arginase	4
Interaction between cytochrome-c and tryptic fragment of cytochrome b-5	5
Dimerization of apo A-II component of high density lipoprotein (low temperature)	3
Aggregation of galactothermin	2
Self plolymerization of S5 subunit of 30S ribosomal unit of *E. coli*	5
Polymerization of S5 with S8 subunits of 30S ribosomal unit of *E. coli*	5
Monomer, dimer, and tetramer association of L7/L12 proteins	4
Ligand binding by proteins	
Copper to serum albumin	2
Certain anions to serum albumin	2
Fatty acid ions to serum albumin	3
Bromophenol blue, bilirubin, and protoporphyrine 1 to the home binding site	4
Interactions involving nucleic acid	
E. coli ribosomal protein S8 binding to 16S r RNA	11
Eco RI endonuclease binding to specific base sequence on plasmed p BR 322 DNA	5
Binding of poly-L-lysine, spermine and Mg^{2+} to DNA	2
Antigen-antibody reactions	
Rabbit antiarsanilic antibody and two haptens	2
Binding of appropriate antibody to antigen-antibody complex where antigens are chemically modified beef pseudoglobulin, beef gamma globulin, and egg albumin	2
Binding of TMV-specific Fab fragments to TMV	3
Binding of ϵ-DNP-L lysine to two IgG preparations	3
Structural and contractile proteins	
Formation of collagen from tropocollagen	2
Formation of myosin filaments from myosin monomer	2
G-actin polymerization to F-actin	2
Interaction of myosin heads with F-actin	2,3
Formation of microtubules from tubulin	2,3
Entropy-driven processes in living cells	
Division of fertilized eggs	2
Formation of pseudopodia in amoebae	2
Protoplasmic streaming in leaf cells of *Elodea canadensis*	2

(Continued)

TABLE 9.1. Entropy-Driven Processes in Biology (Continued)

Process	Reference
Entropy-driven processes in living cells (continued)	
Color changes in killifish	2
Formation of mitotic spindles	2
Formation of axopodia in heliozoa	2
Sickling in sickle cell anemia	2
Development of muscular tension	2

significantly less than TΔS. There are probably many more entropy-driven processes in biology. Search of the literature is difficult because relevant publications can be found readily only if they are indexed in abstracting journals under the heading entropy. Undoubtedly, many are not indexed in this manner. References 2–5 are to a book and reviews in which entropy-driven processes are discussed and original references cited. References 6–11 are to more recent literature.

MUSCLE CONTRACTION

The most extensively investigated entropy-driven process in biology is the contraction of striated muscle. A schematic representation of the structure of a striated muscle cell is shown in Figure 9.1. The muscle itself is a bundle of fibers. Each fiber is a very long multinucleated cell with a membrane known as the sarcolemma. Each fiber is filled with many myofibrils oriented parallel to each other along the long axis of the fiber.

The myofibril is the contractile unit. It is composed of interdigitated thin and thick filaments arranged in parallel. The thin filaments are largely made up of F-actin. The F-actin is a double-stranded helical polymer of globular actin molecules. A complete turn of the helix occurs every 390 Å. Lying in the groove of the actin filament is the protein tropomyosin. Two molecules of tropomyosin occur in every complete turn of the actin helix. In addition, for each turn there are two molecules of troponin bound to the helical actin filament. A schematic representation of the thin filament is shown in Figure 9.2. The thick filaments are longitudinally polymerized myosin molecules. The myosin molecule is a double strand of myosin protein chains in an α helical tail and two heads. This is illustrated in Figure 9.3.

The myofibril contains cross bands that divide it into units called

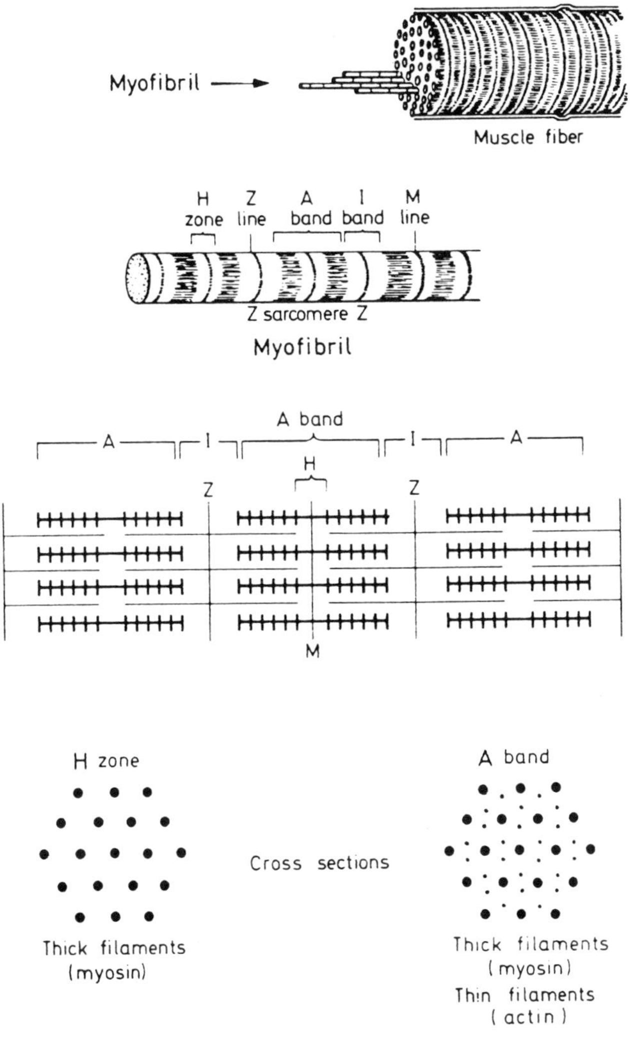

Fig. 9.1. Schematic representation of the structure of the muscle fiber and the myofibril. From Lauffer [2], reprinted with permission of the copyright owner, Springer Verlag.

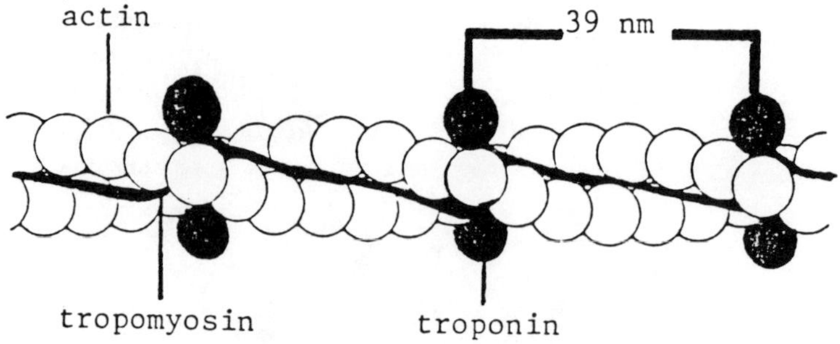

Fig. 9.2: Schematic representation of the thin filament. From Lauffer [2], reprinted with permission of the copyright owner, Springer Verlag.

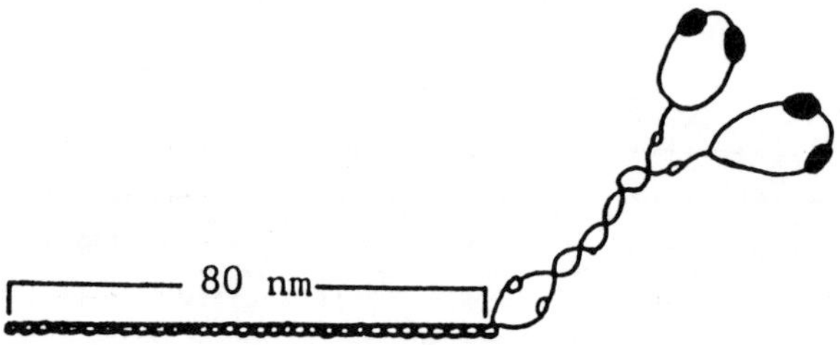

Fig. 9.3. Schematic representation of myosin.

sarcomeres. There are narrow, light bands, known as I bands, with a thin line in the middle known as the Z line. There are dark bands, known as A bands, with a lighter zone in the middle, the H Zone, which in turn is divided by an M line. A sarcomere is the unit between two Z lines. The H zone contains only thick myosin filaments. The I band contains only thin filaments. In the darker portions of the A band, the thick and thin filaments interdigitate. When a muscle contracts, the thick and thin filaments slide together, making the I band and the H zone narrower. The basic chemical reaction of muscle contraction is the combination of myosin heads with actin molecules in the F-actin chains to form actomyosin. Myosin is an enzyme that catalyzes the hydrolysis of ATP. The tropomyosin and troponin of the thin filaments have a regulatory function.

The thin filaments are polar, that is, the actin molecules in the F-actin polymers are all oriented the same way. The polarity is such that within a sarcomere the molecules are oriented in a direction away from the Z

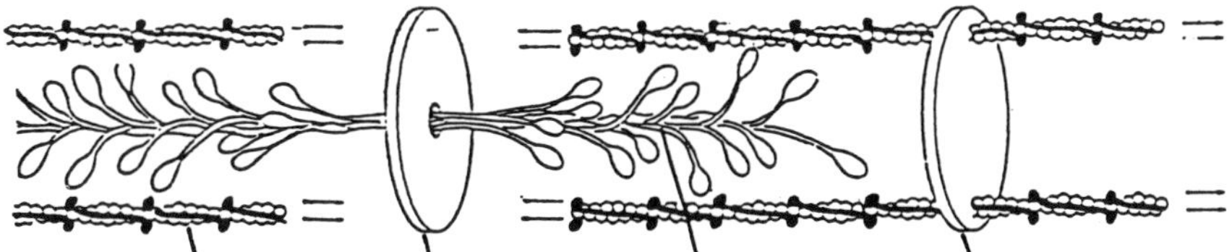

Fig. 9.4. Diagram illustrating the polarity of thick and thin filaments.

line. The Z line contains a protein called α actinin that cements all thin filaments together at this position. The thick filaments are also polar; they are oriented with the myosin heads pointing away from the M line. At the M line, the polarity reverses. There is a protein known as M-line protein, which seems to cement the thick filaments together at that point. These ideas are represented schematically in Figure 9.4.

The kinetics of muscle contraction seems to be exceedingly complex. This is an area under active investigation in many laboratories of the world; many steps involving conformational changes, lever-like actions, etc., have been proposed. Compared with the kinetic picture, the thermodynamic analysis is relatively simple. First, it involves the interaction of actin and myosin heads. In the initial state, the actin and the myosin heads are not combined in relaxed muscle, and in the final state, the myosin heads are combined with actin molecules to form actomyosin in the contracted muscle. The thermodynamics of the interaction between myosin heads and actin units in the F-actin filaments in the absence of ATP has been investigated over a long period of time. References to early work were cited previously by the author [2, 3]. More recently, Yasui et al. [12] have reinvestigated the binding of myosin subfragment-1 with F-actin in the absence of ATP. Subfragment-1 is obtained by enzymatic degradation of myosin; it is the fragment containing myosin heads. The equilibrium constant for the combination of the myosin heads with the F-actin was obtained by measuring the rates of association and of dissociation. Experiments were carried out over the temperature range of 0–38°C. Results in which ln K for association is plotted against reciprocal temperature are shown in Figure 9.5. From these results the standard enthalpy can be determined to be +2.14 kcal/mole of myosin heads and the standard entropy change to be +46.1 cal/deg. mole. This corresponds to a free energy of −12.2

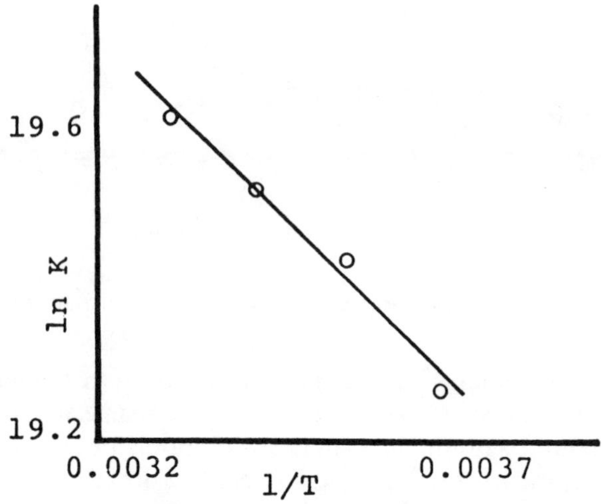

Fig. 9.5. ln K for the binding of myosin heads to F-actin plotted against reciprocal of temperature in Kelvin. Plotted from data of Yasui et al. [12].

kcal/mole at 37°C. Slightly different values of these parameters were published by the original authors [12]. The formation of the actomyosin complex from actin and myosin heads is thus clearly an entropy-driven process. However, the relatively high negative free energy at body temperature shows that this combination is far from reversible. The binding and subsequent hydrolysis of ATP is necessary to reverse this reaction completely. This is consistent with the well-known fact that in death, in the absence of ATP, rigor mortis sets in. Myosin is an actin-activated enzyme that catalyzes the hydrolysis of ATP, thereby making this reversal possible. As discussed in Chapter 7, the standard free energy at pH 7, ΔG^{o}, for the hydrolysis of ATP is -7.3 kcal/mole. However, the actual free energy decrease in the intact living cell is usually much higher than this because of the prevailing concentrations. In human erythrocytes, the molal concentrations of P_i, ADP, and ATP are 1.65, 0.25, and 2.25, all $\times\ 10^{-3}$. When these values are substituted into the equation, $\Delta G = \Delta G^{o\prime} + 2.3RT \log([ADP][P_i]/[ATP])$, one obtains at a temperature of 37°C a value of -12.9 kcal/mole for ΔG. For a muscle that is largely contracted, [AM]/[A][M] should be a large number, considerably greater than 1. Thus, ΔG for the reaction of actin with myosin heads should be significantly smaller than -12 kcal/mole. It is evident, therefore, that the hydrolysis of one molecule of ATP has a sufficiently large negative ΔG to overcome the positive ΔG for dissociating actomyosin into actin and myosin. Thus the net reaction for the dissociation of actomyosin can be written AM + ATP = A + M + ADP + P_i, with a sufficiently large negative free energy change to make the reaction spontaneous. When the concentrations of ATP, ADP, and P_i are all at the appropriate levels, an equilibrium can exist.

$$
\begin{array}{ll}
\mathrm{A + M \rightleftarrows AM} & \Delta G(-) \\
\mathrm{ADP + P_i \rightleftarrows ATP} & \Delta G(+) \\
\hline
\mathrm{A + M + ADP + P_i \rightleftarrows AM + ATP} & \Delta G(0)
\end{array}
$$

Thus, in the forward direction the free energy decrease from the combination of actin and myosin heads can be used to synthesize ATP from ADP and P_i. In the reverse direction, the free energy decrease from the dissociation of ATP can be used to dissociate AM. Many possible equilibria of this sort are postulated by scholars who study the kinetics of muscle contraction. For example, Taylor [13] discusses a complex equilibrium situation involving three conformations of myosin, bound to actin or dissociated from actin, with both free of or bound to the nucleotides ATP, ADP. P_i, and ADP. However, when initial and final states are focused upon, the reaction is as described above. When the muscle is neither contracting nor elongating and in the absence of tension, in the ideal case there is no consumption or buildup of ATP.

From a thermodynamic point of view, the free energy decrease when actin combines with myosin heads during muscle shortening can be used to do work, lifting a weight, for example. In this case, the free energy decrease from the combination of actin and myosin heads is no longer available to synthesize ATP from ADP and P_i. The ATP necessary for breaking the bond between actin and myosin then comes at the expense of the ATP store in the muscle. Thus, in a secondary sense, the energy required when a muscle does work comes from ATP. Phosphocreatine stored in the muscle serves as a reserve donor of high-energy phosphate groups. Phosphocreatine reacts with ADP, catalyzed by the enzyme creatinekinase, to form creatine plus ATP, thus restoring the ATP concentration to its normal value.

Since 1969, the sliding filament-rotating cross-bridge model proposed by Huxley has dominated thinking about the mechanics of muscle contraction [14]. In its simplest form, the cross bridges between the myosin and actin filaments are thought of as being little levers that turn and pull the filaments with respect to one another. In its more

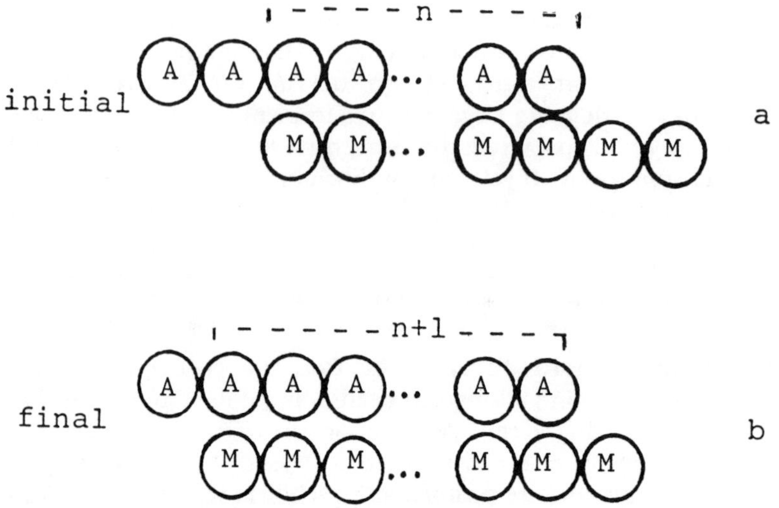

Fig. 9.6. Diagram illustrating one step in the sliding together of thick and thin filaments. A represents the actin subunit and M the myosin head.

modern form, this theory involves conformational change in the myosin heads to pull the filaments with respect to each other. The process is sometimes described as myosin heads "walking" along actin filaments.

Muscle contraction can be understood in terms of the somewhat simplified sequence of reactions described below. Consider a single thin filament and a single thick filament initially in a state in which there are n actomyosin bonds, as illustrated in Figure 9.6a. As was discussed previously, in the absence of ATP each of these actomyosin bonds is formed by an entropy-driven process involving the release of water. To reverse the process and break the bonds, water must be bound. However, under physiological conditions the equilibrium constant is far on the side of actomyosin. To establish true equilibrium under such conditions, something is required in addition. Myosin is an ATPase and thus binds ATP. Thus, to bring the actomyosin bonds in equilibrium, reaction I occurs.

n actomyosin + n ATP $\rightleftarrows$ n actin + n myosin · ATP (reaction I)

Under such equilibrium conditions, the thick filament is able to slide with respect to the thin filament. When the muscle contracts under nonequilibrium conditions, that is, when work is done, energy must be supplied. This comes from the hydrolysis of ATP. First reaction II takes place.

$$\text{n myosin} \cdot \text{ATP} + \text{n } H_2O \rightleftarrows \text{n myosin}' \cdot \text{ADP} \cdot P_i \qquad \text{(reaction II)}$$

In this reaction, associated with the hydrolysis of the bound ATP a conformation change takes place in the myosin to form myosin′, which binds ADP and P_i less strongly than myosin binds them. Now the myosin heads can move to the left one actin unit, and reaction III can take place. These myosin heads do not all move at once; a certain amount of elasticity in the thick filament is required.

$$\text{n myosin}' \cdot \text{ADP} \cdot P_i + \text{n actin} \rightleftarrows \text{n actomyosin}' \cdot \text{ADP} \cdot P_i \qquad \text{(reaction III)}$$

Then comes what is called the power stroke, reaction IV, in which the ADP and the Pi are released and the actin-bound myosin head returns to its original conformation. This change corresponds to the rotation of the "little levers" in the original hypothesis explaining muscle contraction.

$$\text{n actomyosin}' \cdot \text{ADP} \cdot P_i \rightleftarrows \text{n actomyosin} + \text{n ADP} + \text{n } P_i \qquad \text{(reaction IV)}$$

This power stroke brings the final actin unit on the right of the thick filament into contact with a new myosin head, as illustrated in Figure 9.6b. If it is assumed that this myosin head has already bound an ADP and a Pi, then reaction V must take place.

$$\text{actin} + \text{myosin}' \cdot \text{ADP} \cdot P_i \rightleftarrows \text{actomyosin} + \text{ADP} + P_i \qquad \text{(reaction V)}$$

When the conditions are such that the maximum work is done, the net of all five of these reactions is reaction VI.

$$\text{n actomyosin} + \text{actin} + \text{myosin}' \cdot \text{ADP} \cdot P_i + \text{n ATP} + \text{n } H_2O$$
$$\rightleftarrows (\text{n} + 1)\ \text{actomyosin} + (\text{n} + 1)\ \text{ADP} + (\text{n} + 1)\ P_i \qquad \text{(reaction VI)}$$

Whether or not the rotating cross-bridge model is the mechanism of muscle contraction is a matter to be settled experimentally, and extensive attempts to find evidence pro and con are under way. However, when the matter is viewed from a thermodynamic point of view, it is necessary to ask the question as to whether such little levers or their more sophisticated counterparts are necessary. Consider a thick filament in equilibrium with a thin filament in the presence of ATP at a given length of the sarcomere. All of the points of contact between myosin heads and actin molecules are at equilibrium; the bonds are free to make or break without net expenditure of energy. The myosin head at the end of the thick filament can form a new bond with an actin molecule with a very large decrease in free energy. This can happen only if the thick filament advances with respect to the thin one, but since all of the other bonds between actin and myosin are at equilibrium, there is no reason why this should not happen. As soon as the new myosin head-actin bond is formed, the APTase action of the new actomyosin center takes place, and this new bond also comes to equilibrium.

There is nothing new about the conversion of chemical energy into mechanical work without any special machinery. The most commonplace example is the rise of liquid in a capillary. Work is done lifting liquid against gravity, and the sole source of energy is the chemical energy between the molecules of the liquid. A piston can be pushed and do mechanical work when solvent moves by osmosis through a semipermeable membrane into a more concentrated solution. An electrolytic cell connected to a motor converts chemical energy into mechanical energy.

In a moment of idle curiosity, the author performed a little experiment, hitherto unpublished, to demonstrate the direct conversion of chemical into mechanical energy. A straight section of a clear rubber band 1.3 cm in length was attached by a thin wire to one arm of an analytical balance. The weight of the wire + rubber was 0.0337 gm and that of the wire alone, 0.0128 gm. The difference, 0.0209 gm, was the weight of the rubber alone. The band was then soaked for 30 sec in chloroform. Its new weight, wire + rubber · chloroform was 0.0473 gm. Next a vessel containing a layer of chloroform saturated with water under a layer of water saturated with chloroform was raised until the rubber · chloroform was totally submerged in the water layer. The weight of wire + chloroform-soaked rubber, reflecting the buoyancy of the water, was then 0.0166 gm. The water-chloroform vessel was then raised until the tip of the rubber just touched the chloroform at the interface. The weight, including the wire, changed to 0.0353 gm. Thus the pull of the chloroform on the rubber band was 0.0353−0.0166, or 0.0187 gm, or a force of 18.3 dynes. When the vessel was raised further until the rubber band was three-fourths submerged in the chloroform layer, the weight was 0.00842, reflecting the difference between the buoyant action of the chloroform and that of the water. Since the density of the chloroform-soaked rubber is considerably less than that of chloroform, work against gravity was done in submerging the rubber in the chloroform. This was accomplished by a direct conversion of chemical energy into work without the intervention of machinery of any sort. The chemical affinity of chloroform-soaked rubber for chloroform is greater than its affinity for water; no other explanation is required.

Steinberg et al. [15] described a device for the direct conversion of chemical to mechanical work. It involved a belt of the contractile protein, collagen, which moved continuously over a system of pulleys from a concentrated LiBr solution into water. The collagen binds LiBr and contracts in the concentrated salt and dissociates LiBr and relaxes in the water. The diameters of the pulleys are so arranged that this causes rotation and movement of the belt. The energy comes from the transfer of salt from a concentrated to a dilute solution.

The interaction between myosin heads and actin in the myofibril is thought to be prevented by two parallel long tropomyosin molecules per turn of the F-actin helix, located in the grooves of the thin filament. Initiation of muscle contraction is brought about by a shift of the tropomyosin in the groove. This shift in turn is caused by the interaction of calcium ions with the third thin filament protein, troponin. Troponin contains three subunits, a Ca^{2+} binding subunit, an inhibitory subunit, and a tropomyosin binding subunit. Binding of Ca^{2+} alters the interaction between the three subunits of troponin, which in turn alters the interaction between the inhibitory component with actin and tropomyosin and of the tropomyosin binding component with tropomyosin. All of this brings about the shift of the tropomyosin in the groove, thereby strengthening the actin-myosin interaction. Release of the calcium ions that set off this triggering mechanism comes about as a result of a change induced by motor nerve stimulation in the permeability of membranes of tubules, called the sarcoplasmic reticulum, running across the muscle cell. The calcium ion concentration in the cytosol is normally very low, but this nerve-controlled release sets off the mechanism. Muscle relaxation comes about when the nerve impulses cease and the calcium ions are pumped into the cisternae of the sarcoplasmic reticulum. Energy released by ATP hydrolysis operates the pump.

Contractile elements are widespread in living cells other than muscle cells. Contraction in many involves the interaction of nonmuscle myosin with microfilaments composed of actin.

CELLULAR DYNAMICS

Far from being a homogeneous solution, cytoplasm contains, in addition to organelles, an intricate network of microtubules, microfilaments, and intermediate filaments. These structural elements are intimately involved in determining both the shape and the dynamics of the cell.

Microtubules are composed of two kinds of tubulin, α tubulin and β tubulin, assembled in a helical structure into hollow tubes about 18 nm in diameter. Not only are microtubules elements within the cytostructure, but they also form the mitotic spindles, the axoneme of axopodia, and flagella of bacteria and higher organisms. As was reviewed by Lauffer in 1975 [2], the formation of microtubules from tubulin is a reversible, entropy-driven process. The in vitro polymerization is successful in solutions containing ATP or GTP, magnesium ions, and a calcium chelator. Colchicine and calcium ions inhibit polymerization; magnesium ions are essential.

The microfilaments are F-actin, like the thin filaments in the myofibril of muscle. The polymerization of the globular form of actin, G-actin, was reviewed by Lauffer in 1975 [2]. F-actin is a two-stranded helix about 4 nm in diameter in which 27 molecules occupy a length of 72 nm, a full 360° turn. The polymerization of G-actin is an entropy-driven process. Reversible polymerization can take place in the absence of ATP. However, in the presence of ATP, G-actin binds ATP and then polymerizes to form F-ADP actin in the presence of salt. When the salt is dialyzed out, F-ADP actin depolymerizes, but the product is G-ADP actin. In the absence of salt, this reversible transformation can be controlled by changing the temperature. Thus, under conditions of reversibility, ATP is constantly hydrolyzed.

The polymerization of G-actin to form F-actin is often described as a phase transition, the formation of an insoluble complex, and the results of experiments are often analyzed thermodynamically in terms of a critical concentration defined as the concentration of G-actin below which there is no polymerization and that remains constant regardless of the amount of polymerization. It is interesting to consider the relationship of this concept to a simple mass action equation for the reaction

$$nA \rightleftarrows B$$

If this simple equation is to apply, n must have a value of several hundred for G-actin to be in equilibrium with F-actin of any significant chain length. If one neglects activity coefficients one can write

$$K = [B]/[A]^n = (c_b/M_b)/(c_a^n/M_a^n)$$

where c is a concentration in gm/l. This equation can be solved to yield

$$c_a^n = (c_b M_a^n)/(KnM_a)$$

Now, redefine concentration in terms of c', an arbitrary unit chosen such that when $c_b' = c_a'$, $c_a' \equiv 1$. One can then write $c' = \alpha c$. When concentrations are converted to the new arbitrary unit, $(c_a')^n = (c_b')(M_a\alpha)^n/(\alpha KnM_a)$. Since the arbitrary units were chosen such that both c_a' and c_b' are unity when they are equal, the products of the constants multiplying c_b' must be unity. Thus $(c_a')^n = c_b'$ and $c_a' = (c_b')^{1/n}$. When n is infinite, c_a' is equal to unity in the arbitrary unit for all finite values of c_b' because $(c_b')^0$ is unity for all finite values of c_b'. This is the critical concentration in a phase transition. However, even when n is merely a reasonably large number, c_a' calculated from the aforementioned equation varies only slightly when c_b' varies over a wide range. This is shown in Table 9.2. Unless exceptionally accurate analytic procedures are used, it is impossible to distinguish the equilibrium values of c_a' from the true constant it must be if it is the critical concentration in a phase-separation reaction. This is surely the case when n is 100 or more and could be the case when the value of n is considerably smaller than 100. Even when n is as low as 30, the equilibrium values of c_a' vary only $\pm 5\%$ from the mean when c_b' varies over the 25-fold range from 0.2 to 5.0. The concept of critical concentrations is very popular among biochemists, but data of sufficient accuracy to establish with reasonable likelihood that c_a' really is a constant, as required by that concept, are rare.[1]

TABLE 9.2. Values of c_b' and c_a' Compared

				c_a' (n)			
c_b'	10	30	50	100	300	1,500	α^{∞}
.1	0.794	0.926	0.955	0.977	.992	0.997	1.000
.2	.851	0.948	.968	.984	.995	.998	1.000
.5	.933	0.977	.986	.993	.998	.999	1.000
1	1.000	1.000	1.000	1.000	1.000	1.000	1.000
2	1.072	1.023	1.014	1.007	1.002	1.001	1.000
5	1.175	1.055	1.033	1.016	1.005	1.002	1.000
10	1.259	1.080	1.047	1.023	1.008	1.002	1.000
100	1.585	1.166	1.096	1.047	1.015	1.005	1.000

In addition to microtubules and microfilaments, both of which are readily dissociated into their constituent proteins, tubulin and actin, and readily reassembled by entropy-driven processes, there exists a group of relatively stable intermediate filaments with diameters in the region 8–10 nm. Five classes of intermediate filaments have been identified. They are all composed of proteins with high α-helix content and relatively low solubility [16].

FORMATION OF PSEUDOPODIA IN AMOEBAE

Studies carried out by Marsland and associates, discussed in detail by Lauffer [2], show clearly that the ability of amoebae to form pseudopodia, their mechanism of locomotion, depends upon the gel strength of the cytoplasm, in modern understanding, upon the structural network within the cytoplasm. At high pressure and low temperature, amoebae are spherical, but at higher temperatures or lower pressures they exhibit pseudopodia. The formation of the structures that make pseudopodia extension possible is clearly the result of an entropy-driven process, the growth of microfilaments from globular actin, an entropy-driven process in the isolated state. Thus cell shape is demonstrated to be related to structures generated by entropy-driven processes. The cytoplasm of the amoeba, as illustrated diagrammatically in Figure 9.7, consists of a central core, called endoplasm or plasma sol, and a relatively thick cortical layer surrounding it, called ectoplasm or plasma gel [17]. When a pseudopodium forms, the endoplasm streams in the direction of extension and at the tip congeals into gel-like ectoplasm. In other regions of the cell, ectoplasm changes into plasma sol. Thus,

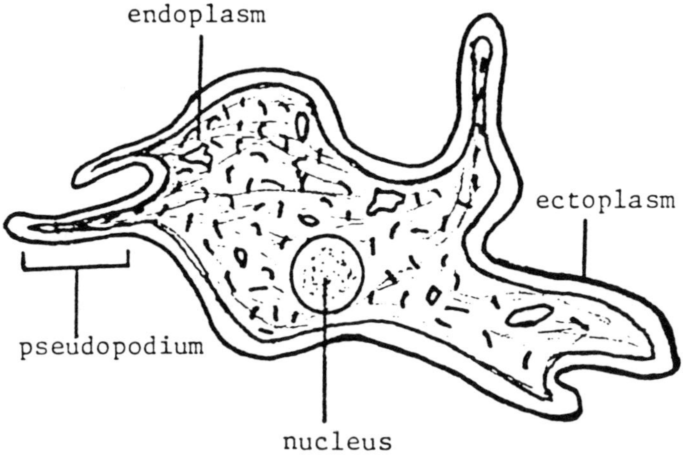

Fig. 9.7. Idealized representation of a moving amoeba.

pseudopodia formation is a reversible sol-gel transformation, a G-actin-F-actin transformation, long known to be an entropy-driven process. The net result is a displacement of the center of mass in the direction of the new pseudopodium. Thus the amoeba moves. Energy is supplied by the hydrolysis of ATP.

CELL DIVISION

Early work of Marsland and associates, discussed in detail by Lauffer [2], shows that the ability of fertilized eggs to divide is correlated with cytoplasmic structure formed by an entropy-driven process. When a fertilized egg divides, there is a gradual change from a single large sphere into two smaller spheres, at first partially attached and later separated. There is thus a change in shape that cannot happen unless reversible structure-forming processes operate. In addition, karyokinesis must occur. Inoué, as reviewed by Lauffer [2], demonstrated quite clearly that the spindles responsible for chromosome separation are formed by an entropy-driven process. Evidence reviewed by Euteneuer [18] supports the hypothesis that the microtubules of which the spindles are composed are polar, with oppositely oriented tubules originating at the opposite poles in the spindle. At the beginning of chromosome separation, these oppositely oriented microtubules are interdigitated within the spindle, their distal ends attached to chromosomes. As division proceeds, these oppositely oriented microtubules slide apart, pulling the attached chromosomes with them. While some theories of chromosome separation involve force generation within the interaction of the oppositely oriented microtubules, a difficulty to that theory is known. As the microtubules slide apart, the spindle of necessity elongates. The maximum possible elongation from this source alone is to double the separation of the poles. In some species, however, the poles separate more than double their original separation. Thus, if force generation does exist between these microtubules, there must be additional force generation external to the region of overlap. As of 1986, the question of force generation in karyokinesis was still unsettled.

The sliding of microtubules with respect to one another is also involved in the motion of cilia and flagella of eukaryotic cells. This motion has been extensively investigated. The energy comes from the hydrolysis of ATP catalyzed by an ATPase, dynein. This enzyme appears in electromicrographs as crossarms between microtubule doublets. A similar mechanism presumably exists for the movement of microtubules during kariokinesis. However, there are two distinct differences between these two cases. First, in cilia the sliding tubules are oriented parallel, not opposite as in spindles. Second, even though electron micrographs reveal some cross-bridging material between the tubules in the spindle, antibody-staining evidence shows that it is not dynein. It is, therefore, postulated that in spindles there is a dynein-like ATPase that mediates tubule separation. What kind of mechanism could account for the mediation of the ATPase in the separation of microtubules? Final answers are not yet available; speculation is all that is available. First of all, it must be remembered that the microtubules being separated are oppositely oriented. A simple mechanism, if dyneinlike molecules were free to rotate, would be one analogous to a rack and pinion, where the dyneinlike molecule is the counterpart of the pinion. Another possible mechanism would be one in which the ATPase is a tail-to-tail dimer with each head oriented in the same direction with respect to the orientation of one of the oppositely oriented microtubules. Then "forward creeping" of each head along its microtubule track would force the two tubules to move in opposite directions. ATP hydrolysis would furnish the energy.

As cell division proceeds, the mother cell divides in two, isolating the chromosomes separated by kariokinesis into two daughter cells. This division is brought about by the contraction of a contractile ring, whose plane is perpendicular to the spindle axis. The contractile ring is composed largely of actin filaments associated with myosin, and its contraction comes about by the interaction of myosin and actin. Like the spindles, the contractile rings are temporary structures that appear during cell division and disappear when no longer needed. As previously noted, the polymerization of tubulin into microtubules and the formation of F-actin from G-actin are both entropy-driven processes, readily reversible under physiological conditions.

PIGMENT MIGRATION

Pigment migration in melanophores of the killifish was shown by the studies of Marsland and associates, as reviewed by Lauffer [2], to depend upon the existence of structures formed by an entropy-driven process. Euteneuer [18] reviewed evidence showing that microtubules in the arms of the melanophores are all of the same polarity. High pressures and low temperatures favor pigment dispersal, whereas high temperatures and low pressures, conditions favorable for microtubule growth, favor aggregation of the pigment granules in the central region of the cell. Under these conditions, the killifish appears pale, but when the granules are dispersed, they give the fish a dark color.

CYTOPLASMIC STREAMING

As was shown in Table 9.1, protoplasmic streaming has long been known as an entropy-driven process. Considerable detailed information is available concerning the cytoplasmic streaming in giant algal cells [19]. An endless ribbon of cytoplasm streams in a helical path up one side of the cell and down the other, dragging the organelles along. The approximate structure of such cells involves a stationary layer of chloroplasts immediately adjacent to the cell wall. Immediately inside the layer of chloroplasts is a subcortical layer of actin filaments. Then comes a thin layer of moving cytoplasms and in the center of the cell is a large vacuole. As in striated muscle, the actin filaments are oriented unidirectionally along the continuous path of the streaming.

Porter [20] reviewed evidence concerning the relationship of protoplasmic streaming to actin filaments in the endoplasm. Suggestions of the way in which these microfilaments generate motion include the possibility that the filaments execute an undulating motion and the possibility that some myosinlike protein moves along fixed actin fibers. Porter concludes that whatever the source of the motive force, the involvement of actin filaments in cytoplasmic streaming seems to be incontrovertible. Today the widely held explanation for the streaming is that plant myosin is attached to organelles in the cytoplasm, the heads of which interact with the stationary actin filaments, moving along the filaments in one direction only. This motion exerts a viscous drag on the cytoplasm, in some ways analogous to the drag that moving ions exert on the liquid in electroosmosis. Velocities of streaming have been observed as high as 75 μm per second.

The thermodynamic explanation of this movement is somewhat like that in contraction of muscle. When a myosin head attaches to an actin monomer, there is a decrease in free energy by an entropy-driven process. The binding of ATP breaks the actomyosin bond, and its hydrolysis prepares the way for the myosin head to move on and attach to the next actin subunit.

It is easy to understand from a thermodynamic point of view why muscle contraction is unidirectional. Only by the sliding of the thick myosin filaments and the thin actin filaments into each other can the chemical reaction between actin and myosin proceed in a forward direction. However, in the case of cytoplasmic streaming one must ask why when a bond between a myosin head and an actin subunit is broken by ATP binding does the myosin head attach itself to the next actin subunit in front of it rather than, with equal probability, to the unit behind it. A purely mechanical contribution involves inertia. Once the cytoplasm starts to stream, then, at the instant that a myosin head becomes free, the viscous drag of the moving fluid on the organelle attached to the plant myosin unit will urge it forward to bring it closer to the actin subunit in front of it than to that behind it. Then the interaction of this head with the new actin subunit will give an additional surge in the forward direction. This would be a sufficient explanation for multidirectional streaming, as found in many plant cells, for streaming that is interrupted could start again in either direction.

If streaming always is in the same direction, as it seems to be in the giant algae, then some additional factor must be involved. When more than simple inertia is required to explain the uniform forward movement of the plant-myosin heads along the actin filaments, conformational changes in the myosin can be invoked. As was discussed previously, students of the interaction of actin and myosin in muscles postulate that the actomyosin ATP interaction involves a conformational change in the myosin head in a manner that creates a unidirectional stress on the myosin tail. The stress is relieved by altering the position

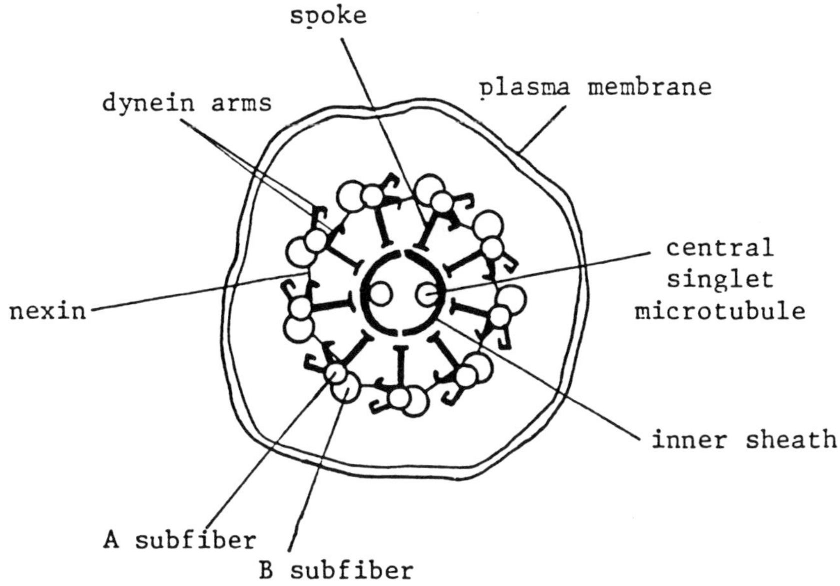

Fig. 9.8. Cross-sectional diagram of a cilium.

of the myosin head with respect to monomeric units in the actin chain. If this is indeed the way muscle contraction works, then if a similar mechanism is involved in the interaction of plant-myosin with actin filaments, movement of the myosin and the attached organelles should not only be unidirectional but should always be in the same direction.

MOVEMENT OF CILIA

Cilia are small hairlike protrusions about 0.25 μm in diameter that extend from the surface of animal and some plant cells. Their principal function is to move fluids past the cell surface. The base of each cilium is firmly attached to the cell membrane. A typical cilium is about 10 μm in length. They beat as a result of a bending motion. The flagella of many protozoa and of sperm, unlike those of bacteria, behave like very long cilia and derive their propelling force from a similar bending motion. A cross-sectional diagram of a cilium is shown in Figure 9.8. It has long been known that in cross section cilia exhibit a 9 + 2 arrangement. Nine doublet microtubules are arranged in a circle surrounding two

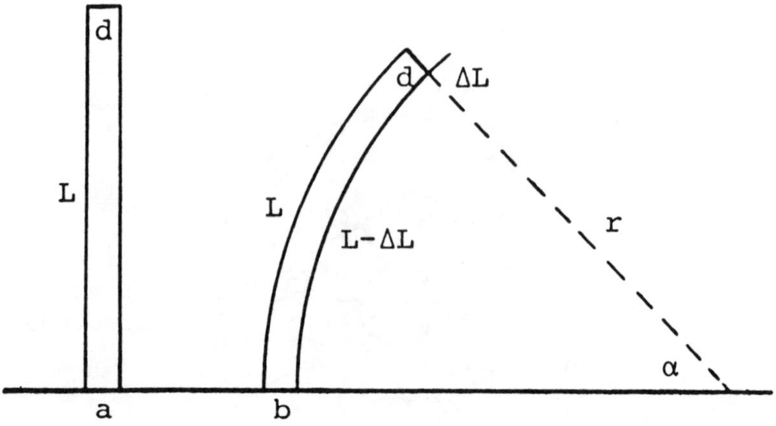

Fig. 9.9. Idealized diagram showing how sliding of one thin filament with respect to another causes bending of the cilium or flagella. The angle α is the angle subtended by the arc and also the angle between the tangent of the cilium at its end and the original vertical position.

central singlet microtubules. Surrounding the pair of central singlets is an inner sheath. Radial spokes extend from the inner sheaths to each of the nine doublets. Each doublet is composed of an A subfiber extending the entire length of the cilium and a B subfiber that is a little shorter. The A subfiber is a complete cylinder with 13 tubulin molecules in a single turn. The B subfiber is an incomplete cylinder with ten tubulin molecules and is attached to the A subfiber in such a way that part of the wall of the A subfiber is shared with the B subfiber. Connecting one doublet with its neighbor is an elastic protein, nexin. Also extending from the A subfiber in a clockwise direction toward the B subfiber of the neighboring doublet are two arms composed of dynein, a protein that has ATPase activity. The bending of the cilium comes about as a result of the sliding of the doublets with respect to one another. The elastic nexin prevents this sliding from going too far. Just how such sliding motion can produce bending is illustrated diagrammatically in Figure 9.9 This is an imaginary structure of exagerated proportions with only two diametrically opposed doublets of equal length, shown straight (a) and bent to the right (b). The radius of curvature is represented by r and the angle subtended by the arc as α. If L is the length of each of the doublets and d is the distance of separation of the two doublets, then for the distal doublet, $L = 2\pi(r + d)\,\alpha/360$. For the proximal doublet, L −

$\Delta L = 2\pi r\alpha/360$. By subtraction, $\Delta L = 2\pi d\alpha/360$. Thus, when the hypothetical structure bends so as to subtend an angle α, the proximal doublet must slide toward the tip by a distance ΔL with respect to the distal doublet. The angle, α, is also the angle between a tangent at the tip of the bent cilium and the original vertical direction. In the actual cilium, the doublets are not diametrically opposed as in this idealized diagram, but the principle is the same; doublets on the side nearer the direction of bending slide outward with respect to their more distal neighbors. This motion is generated by the movement of the dynein arms extending from the A subfiber of a more distal doublet along the B subfiber of its more proximal neighbor. Dynein is an ATPase requiring Mg^{2+}. The energy for this movement comes from the enzymatic hydrolysis of ATP. The evidence supporting this analysis of the motion of cilia was reviewed by Satir [21].

There are striking similarities and dissimilarities, discussed in two reviews by Johnson and collaborators [22, 23], between the dynein-microtubule system and the much better understood myosin-filamentous actin system. Both myosin and dynein are ATPases. While myosin has two heads attached to a long rod, dynein has three heads attached to a very short rod. The structure of dynein is sometimes described as a bouquet in which the three heads are the buds and the short rod coresponds to the stems. In both cases, the enzymatic activity resides in the heads. In both cases, the combining sites are also in the heads. A major difference is that in the sliding filaments of cilia, the stems of the dynein are firmly attached to the A subfiber of one doublet, and the heads interact with the B subfiber of a neighboring doublet, in contrast with the structure in skeletal muscle already discussed. In both cases, the role of ATP is to dissociate the bonds between the head and the appropriate filament. The mechanism involved in doublet sliding is assumed to be comparable with that of the well-investigated sliding of thick versus thin filaments in muscle. Similar reaction schemes are postulated. Holzbauer and Johnson [24] measured the binding constant of dynein for ATP by dynein and from that measurement calculated that at a concentration of 1 mM ATP, the free energy change for ATP binding is $\Delta G = -3.25$ kcal/mole. They conclude, "in the cross-bridge model of the dynein-microtubule cycle, ATP binding has been shown to induce the release of the dynein heads from the microtubules, and ATP hydrolysis then occurs on the free molecule. Although the dynein-microtubule binding equilibrium has not yet been determined, it is clear that the large free energy change accompanying ATP binding is sufficient to drive the dissociation of dynein from the microtubule. . . . "

An additional similarity between the two systems is that both actin thin filaments and microtubules are produced by the polymerization of their constituent monomers by entropy-driven processes. In the actin-myosin system, the interaction between myosin heads and filamentous actin is also known to be an entropy-driven process. However, as was just mentioned, in the dynein-microtubule system, thermodynamic data on the interaction of dynein and microtubules are not available. One can only speculate. The bond between dynein and the microtubules must, with the aid of ATP binding, be readily reversible at biological temperatures. As was pointed out in Chapter 8, entropy-driven processes are admirably adapted to easy reversibility without sacrificing the actual strength of the bonds. By this reasoning, one would expect the dynein-microtubule bond to be one formed by an entropy-driven process. However, as the hypothetical numbers in Table 8.1 of Chapter 8 illustrate, bonds formed by enthalpic reactions can also reverse at biological temperature, provided that they are weak enough. If sufficiently weak bonds between dynein heads and microtubules suffice for the motion of cilia, eucaryotic flagella, and spindle elongation in cariokinesis, then it might turn out that the dynein-microtubule interaction is not entropy-driven. The answer will come only by experimentation.

BACTERIAL MOTILITY

Many bacteria swim as a result of the motion of a flagellum. Unlike the situation with respect to cilia and flagella of eukaryotic organisms, the bacterial flagella move not as a result of sliding of filaments, but by actual rotation. The flagellum is the analog of the propeller in a motorboat. The bacterium contains a motor consisting, in the idealized case, of a stationary element attached to the cell surface, a rotating element attached to the flagellum, and a bearing attached to the outer membrane. Figure 9.10 is a speculative model. That the flagellum actually rotates can be demonstrated by attaching the flagellum to a solid surface and observing the rotation of the bacterium.

The source of energy for the operation of the flagellar motor is not the hydrolysis of ATP, at least not the immediate source. Rather, it is the flow of protons across a pH gradient. Convincing evidence of this is that

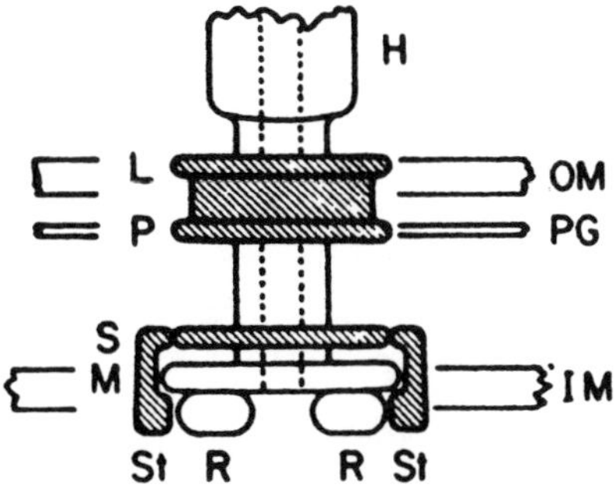

Fig. 9.10. A speculative model for the structure of the bacterial flagellar motor. Stationary features are crosshatched. Reprinted from Macnab and Aizawa [25], with permission of the copyright owner, Annual Reviews, Inc.

the motor can be operated in the absence of ATP but by the imposition of an artificial pH gradient. Mounting evidence suggests that the device works with a fixed stoichiometry of protons per revolution.

In Chapter 7 it was mentioned that bacterial, chloroplast, and mitochondrial membranes contain a system involving a channel for H^+ ions designated F_0 and, on the inside, an ATPase designated F_1, involved in the synthesis of ATP. This system was discussed in Chapter 7 as a device for movement of H^+ ions against a concentration gradient, with energy supplied by the hydrolysis of ATP. However, it was originally proposed by Mitchell in connection with the synthesis of ATP with the motive force being the flow of protons down a concentration gradient. The source of the hydrogen ion gradient is the electron transport system in the metabolic process. Mitchell proposed in 1961 that electron transport is accompanied by a net movement of hydrogen ions from the inside to the outside of the inner mitochondrial membrane. A proton motive force (pmf), resulting in part from the pH gradient across the membrane and in part from the membrane potential, is the actual immediate energy source.

$$\Delta(\text{pmf}) = \Delta \mathbf{V} - 2.3\ \text{RT}\ \Delta\ \text{pH/F}$$

How can such a machine work? A plausible analog is a water wheel in which water captured by the cups on the wheel at a high elevation, therefore at high potential energy, can be released at a low elevation, therefore at a lower potential energy, only by the rotation of the wheel.

In the case of a flagellar motor, a proton binding source on the rotor in one position when it is in contact with the side of the membrane at high hydrogen ion concentration must move in the direction of rotation to bring it in contact with the other side of the membrane where the hydrogen ion concentration is lower. Details of the structure of the flagellar motor are under active investigation. The status as of 1984 was reviewed by Macnab and Aizawa [25].

REFERENCES

1. Alberts, B., Bray, D., Lewis, J., Raff, M., Roberts, K., and Watson, J.D. Molecular Biology of the Cell. New York and London: Garland Publishing, Inc., 1983.
2. Lauffer, M.A. Entropy-Driven Processes in Biology. Berlin: Springer-Verlag, 1975.
3. Lauffer, M.A. In Physical Aspects of Protein Interactions, Catsimpoolas, N. (ed.). Amsterdam: Elsevier North Holland, 1978 and Dev. Biochem. *3:*115–170, 1978.
4. Lauffer, M.A. Comments Mol. Cell. Biophys. *2:*99–109, 1983.
5. Shalaby, R.A. and Lauffer, M.A. Arch. Biochem, Biophys. *236:*390–398, 1985.
6. Menegatti, E., Guarneri, M., Bolognesi, M., Ascenzi, P., and Amiconi, G. J. Mol. Biol. *191:*295–297, 1986.
7. Ascenzi, P., Amiconi, G., Bolognesi, M., Menegatti, E., and Guarneri, M. Biochem. Biophys. Acta *832:*378–382, 1985.
8. Markina, V.L., Sudzubiene, O., Pesliaskas, J., and Metelista, D.I. Biokhimiya (Moscow) *51:*306–312 (Chemical Abstracts 104, 164190d, 1986), 1986.
9. Lewis, M.S. and Youle, R.J. J. Biol. Chem. *261:*11571–11577, 1986.
10. Luther, M.A., Cai, G.-Z., and Lee, J.C. Biochemistry *25:*7931–7937, 1986.
11. Mougel, M., Ehresmann, B., and Ehresmann, C. Biochemistry *25:*2756–2764, 1986.
12. Yasui, M., Arata, T., and Inoue, A. J. Biochem. (Tokyo) *96:*1673–1680, 1984.
13. Taylor, E.W. Comments Mol. Cell. Biophys. *1:*115–128, 1981.
14. Huxley, H.E. Science *164:*1356–1966, 1969.
15. Steinberg, I.Z., Oplatka, A., and Katchalski, A. Nature *210:*568–571, 1966.
16. Geisler, N. and Weber, K. In Cell and Molecular Biology of the Cytoskeleton, J.W. Shay (ed.). New York: Plenum Press, 1986, pp. 41–68.
17. Taylor, D.L. and Condeelis, J.S. Int. Rev. Cytol. *56:*57–144, 1979.
18. Euteneuer, U. In Cell and Molecular Biology of the Cytoskeleton, J.W. Shay (ed.). New York: Plenum Press, 1986, pp. 179–202.
19. Allen, N.S. Can. J. Bot. *58:*786–796, 1980.
20. Porter, K.R. In Cell Motility, R. Goldman, T. Pollard, and J. Rosenbaum (eds.). Coldspring Harbor Conferences on Cell Proliferation, Vol. 3, Book A, 1976, pp. 1–28.
21. Satir, P. Sci. Am. *231:*44, 1974.
22. Johnson, K.A. Annu. Rev. Biophys. and Biophys. Chem. *14:*161–188, 1985.
23. Johnson, K.A., Marchese-Ragona, S.P., Clutter, D.B., Holzbaur, E.L.F., and Chilcote, T.J. J. Cell Sci *Suppl 5:*189–196, 1986.

24. Holzbaur, E.L.F. and Johnson, K.A. Biochemistry *25*:428–434, 1986.
25. Macnab, R.M. and Aizawa, S.I. Annu. Rev. Biophys. Bioeng. *13*:51–83, 1984.

NOTES

1. The main reason for determing the equilibrium constant is to evaluate $\Delta G^\circ = -RT \ln K$. For the reaction, $nA \rightleftarrows B$, K is $[B]/[A]^n \cdot K$ and the ΔG° calculated from it are for the formation of 1 mole of B. The useful ΔG° value is the average value per bond, which is $1/(n - 1)$ times the ΔG° value thus calculated. When n is unknown, however, these equations cannot be solved exactly.

An approximate solution, which yields good accuracy when $n > 100$ follow from considering the reactions, $A + A \rightleftarrows A_2$; $K = [A_2]/[A]^2$; $A_2 + A \rightleftarrows A_3$; $K = [A_3]/[A]_2[A]$; $K^2 = [A_3]/[A]^3$. By extending the argument, $K^{n-1} = [A_n]/[A]^n$. The above deduction involves the assumption that the value of K is the same for every new bond formed. $K^{(n-1)/n} = [A_n]^{1/n}/[A]$. As $n \rightarrow \infty$, $K = 1/[A]$ and $\Delta G^\circ = -RT \ln 1/[A]$. This is the same as the equation for a phase transition. Even when n is as low as 100, this equation yields a sufficiently accurate approximation of ΔG° for the formation of one bond. Thus, from a practical point of view, considering the reaction as a phase transition, the formation of an insoluble product from soluble reactants, and [A] as a critical concentration leads to a useful result, even though, from a philosophical point of view, the reaction is actually the formation of a soluble polymer from n monomers.

Index

Abramson, H.A., 116, 125, 126
Actin filaments, 224
cross-bridges with myosin, 229-230
in endoplasm, and cytoplasmic streaming, 239-240
interaction with myosin, 227-229, 243-244
in cell division, 239
in cytoplasmic streaming, 240
equilibrium constant in, 227-228, 247
regulation by troponin and tropomyosin, 233
microfilaments in cytoplasm, 234
polarity of, 226-227
Activity coefficient
of electrolytes, 99-103
in liquid junction potentials, 154, 156
and osmotic pressure, 14, 19
Actomyosin, 227, 228
dissociation of, 229
Aizawa, S.I., 245
Alberts, B., 221
Albumin, bovine serum, and polymerization of tobacco mosaic virus protein, 212
Amoebae, pseudopodia formation in, 236-237
Anomalous osmosis, 173-177
Ansevin, A.T., 197
Antitransport, 181
Arnold, H.D., 37
Arrhenius, S.
on flow of liquids, 26
method for degree of dissociation, 105
Association-induction hypothesis, and unequal ionic concentrations inside and outside of cells, 170
ATP
and active transport across membranes, 183-184
hydrolysis of, 169, 173
catalyzed by dynein, 238
and dissociation of actomyosin, 228-229
thermodynamics of, 181-183
ATPase
calcium, 183-184
sodium-potassium, 169, 186

Bacterial motility, 244-246
Bacteriophage adsorption to bacteria, 72-73
Banerjee, K., 197, 198, 208
Benzinger, T., 182
Bessel functions, 74
of second kind, modified, 133, 134
Beutner, R., 164-165
Bier, M., 116, 130
Bingham, E.C., 45
Bloomer, A.C., 211, 215
Boltzmann distribution law, 92, 95
Booth, F., 129, 130, 138
Briggs, D.R., 51, 124
Bull, H.B., 28, 46, 51, 109, 122, 125, 140-142, 181, 217
Butler, T.J.G., 211

Calcium ATPase, 183
Calcium ions
binding in polymerization of tobacco mosaic virus protein, 213-214
release in muscle contraction, 233
Calomel electrodes, 152
and Donnan potentials, 166-167
and liquid junction potentials, 160-161
and membrane potentials, 163-164
Capillaries
flow through, 25, 27-31
electroosmotic, 119
in lung, carbon dioxide transport in, 181
streaming potential in, 122-126
Casper, D.L.D., 197, 198, 211
Cell division, 237-239
Cellular dynamics, 234-236

Centimeter-gram-second system compared to m.k.s. system, 89–90, 91
c.g.s. system compared to m.k.s. system, 89–90, 91
Charge
  electronic, determinations of, 38
  and electrophoretic mobility, 140
  opposite. *See* Opposite charges
  per unit area
    of flat surface, 132–133, 143–144
    of sphere, 130–132
Charged macromolecules
  diffusion of, 80–81
  and electroviscous effect, 51
  and equation for osmotic pressure, 16
  hydration of, 14
  and osmotic pressure in solutions, 9–13
Charged particles
  spherical, and Debye-Hückel theory, 94–99
Charged spheres, motion of, electrostatic forces in, 108
Chemical energy, conversion to mechanical work, 232–233
Chloride concentrations, inside and outside of cells, 168–169
Chromosome separation, mechanism in, 237–238
Cilia motion, 241–244
  sliding of microtubules in, 238, 243–244
Cohn, E.J., 212
Concentration cells
  with liquid junction, 158–161
  without transference, 156–158
Conductivity
  and diffusion of charged macromolecules, 80–81
  of electrolyte solutions, 103–105
  equivalent, 104, 108
    and liquid junction potentials, 155
  specific, 104
    and streaming potential, 124–125
Contraction of striated muscle, as entropy-driven process, 224–234
Copper electrode in Daniell cell, 147–148
Coulomb's law, 90
Counter charges. *See* Opposite charges
Crank, J., 74, 75
Cresol, separating sodium chloride and potassium chloride, 165
Curran, P.F., 82
Cylindrical diffusion, radially symmetrical, 74–79, 87–88
Cylindrical surface, charge on, 133–138, 144–145
Cytoplasm, structural elements in
  and cell division, 237–239
  and cellular dynamics, 234–236
  and pseudopodia formation in amoebae, 236–237
Cytoplasmic streaming, 239–241

Daniell cells, 147–150
Debye-Hückel approximation, 96, 116
Debye-Hückel constants, 96
  for aqueous solutions, 102
Debye-Hückel theory, 94–99, 101, 103, 114
  and charge per unit of sphere, 130–132
Definite integral, evaluation of, 66, 85–86
De Vries, H., 1-2
Diffusion, 59–88
  of charged macromolecules, 80–81
  from concentrated to dilute solution, 160
  concentration-dependent, 79–80
  and estimation of volume fraction of solute, 70
  facilitated, 179–181
  Fick's law of. *See* Fick's law of diffusion
  in gels, 64–66
    obstruction effect in, 53, 64
  in obstructed system, 79
  osmotic pressure in, 2, 59
  radially symmetrical, cylindrical, 74–79, 87–88
    relevance to biology, 75–78
  radially symmetrical, spherical, 70–74, 86–87
    relevance to biology, 72–74
  of salts, 111–114
  thermal, 81–83
  through pores, 62–64
  and transport across membranes, 177–181
  viscosity in, 48
Diffusion-limited reaction rates, 73–74
Diffusion potential
  and liquid junction potential, 160

and transport in absence of membranes, 161
Dilatometer of Linderstrøm-Lang, 194
Dissociation
of actomyosin, 183, 228–229
of electrolytes, 105
Donnan equilibrium, 166, 168
and osmotic pressure, 9–13
Donnan potentials, 166–167, 168
Donnan term, 13, 16–18
Dorn effect, 115
Durham, A.C.H., 197, 211
Dynein
and ATP hydrolysis, 238
and cilia movement, 242–244
structure of, 243

Ectoplasm, and pseudopodia formation in amoebae, 236
Einstein equation for viscosity of suspension of spheres, 23, 44–45, 52
Einstein-Sutherland equation, 60, 73, 83, 111
Electrical neutrality. *See* Neutrality, electrical
Electrical work in polymerization of tobacco mosaic virus protein, 201–202, 205–207, 220
Electric fields, motion in, 89–114
and activity coefficient of electrolytes, 99–103
and Boltzmann distribution law, 92
and comparison of c.g.s. and m.k.s. systems, 89–90, 91
and conductivity of electrolyte solutions, 103–105
and Debye-Hučkel theory, 94–99
and diffusion of salts, 111–114
and electrophoretic effect, 105–107, 129
and ionic mobility, 108–111
and Poisson's equation, 92–94
and strength inside and outside of sphere, 97
and time of relaxation effect, 108, 129, 130
Electrode potentials, 147–152
standard reduction, 149–150
Electrokinetics, 115–146
charge per unit area
of cylindrical surface, 133–138, 144–145
of flat surface, 132–133, 143–144
of sphere, 130–132, 143
electroosmosis, 115, 116–121
electrophoresis, 115, 121–126
streaming potential, 115, 121–126
theory verification, 138–142
Electrolytes
activity coefficient of, 99–103
concentrations of
and activity coefficient, 103
and conductance, 104, 105
and diffusion coefficients, 114
and electroviscous effect, 51
conductivity of solutions, 103–105
equivalent, 104, 108
specific, 104
dissociation of, 105
and Donnan effect, 17–18
and hydration of macromolecules, 14
ionized, osmotic pressure of, 5
Electromotive force, 11
in Daniell cell, 148–149, 171
Electronic charge, determinations of, 38
Electroosmosis, 115, 116–121
and anomalous osmosis, 174
and streaming potential, 125
Electrophoresis, 115, 126–130
and charge related to mobility, 140
and ionic strength affecting mobility, 140
moving boundary method in, 109
theory verification, 138–142
Electrophoretic retardation effect, 105–107, 129
Electrostatic interactions, and ideality of ions, 101
Electrostriction, 109–110
Electroviscous effect, 51
Ellipsoids of revolution
intrinsic viscosity of, 47–48
oblate or flattened, 381, 39
friction coefficient for, 40–41, 57
prolate or elongated, 38, 39
friction coefficient for, 39–40, 56–57
Endocytosis, 186–187
Endoplasm, and pseudopodia formation in amoebae, 236
Endothermic processes, 191
Energy barrier, potential, and flow of liquid, 27

England, A., Jr., 212
Enthalpy
 compared to entropy-driven processes, 215
 equation for, 190
 for hydrolysis of ATP, 182–183
 and osmotic pressure, 8, 15–16
Enthalpy change
 negative, 191
 positive, 191
Entropy-driven processes, 189–195
 biological applications of, 221-247
 in cell division, 237-239
 in cellular dynamics, 234–236
 in cilia motion, 241-244
 compared to enthalpy-driven processes, 215
 in cytoplasmic streaming, 239-241
 examples of, 222–224
 and hydrolysis of ATP, 182–183
 in muscle contraction, 224–234
 and osmotic pressure, 6–8
 in pigment migration, 239
 in polymerization of tobacco mosaic virus protein, 195–215
 in pseudopodia formation in amoebae, 236–237
 significance in biology, 215–219
Equilibrium Donnan, 166, 168
 and osmotic pressure, 9–13
Equilibrium constant
 hydrostatic pressure affecting, 192–193
 in myosin interaction with actin, 227–228, 247
 temperature affecting, 193–194
Equilibrium potentials, 165–166
Error function method in solution of Fick's second law, 69
Esters streamed through porous cellulose diaphragms, 125–126
Euteneuer, U., 237, 239
Excluded volume, 73
 and osmotic pressure, 5–9
 and transport in gels, 65
Exocytosis, 186–187
Exothermic processes, 191
Eyring, H., 26–27, 50–51

Fick's law of diffusion, 59–62
 first law, 60–61, 63, 82, 83
  diffusion coefficient in, 84
  for gels, 64
  for radially symmetrical cylindrical diffusion, 74
  for radially symmetrical spherical diffusion, 70
 second law, 62
  diffusion coefficient in, 84
  for gels, 65
  for radially symmetrical cylindrical diffusion, 74
  for radially symmetrical spherical diffusion, 71
  solutions of, 66–70, 84–85
Flagella motion, 241
 in bacteria, 244–246
 sliding of microtubules in, 238
Flow of liquids. *See* Hydrodynamics
Fluidity as reciprocal of viscosity, 45–46, 58
Fourier, J.B., 60
Fraenkel-Conrat, H., 211
Free energy
 and ATP hydrolysis, 181–183
 Gibbs
  in liquid junction potentials, 153
  in polymerization of tobacco mosaic virus protein, 195
Freundlich, H., 115
Fricke, H., 53
Friction
 and diffusion, 79
 rotational, 31–35
Friction coefficient
 for ellipsoids of revolution, 38–41, 54–57
 for random orientation, 39, 54–56
  of cylinders, 138
Frictional resistance, 23–58. *See also* Resistance, frictional

Gallagher, W.H., 212, 213
Gans R., 39
Gas law, ideal, 92
Gas molecules, distribution in gravitational field, 92
Gauss' law, 90, 92, 144
Gels, diffusion in, 64–66
Gibbs-Duhem equations, 2–4
Gibbs free energy

in charged particles, 102
in liquid unction potentials, 153
in polymerization of tobacco mosaic virus protein, 195
Glass electrodes, operation of, 164
Glasstone, S., 2
Glucose 6-phosphate, 183
Glucose transport, carrier-mediated, 181
Glutamyl transferase, 183
Glycerin interaction with water molecules, 111
Glycine, and polymerization of tobacco mosaic virus protein, 212
Goldman equation, 169
Gorin, M.H., 116, 133
Gortner, R.A., 15, 21, 142
Gouy, L., 116
Group translocation, 184
Güntelberg, E., 102

Hagenbach, E., 28
Hamburger, H., 2
Hankel functions, 133, 134
Hearst, J.E., 15, 21
Helmlholtz, H., 94, 116
Helmholtz double layer, 94, 116
Helmholtz-Smoluchowski equation, 128
Henderson integration, in liquid junction potentials, 154
Henry's equation, 128–129, 138
Henry's law, 99
Herzog, R.O., 39
Hexokinase, 183
Hill, A.V., 15, 21
Hill, T.L., 133
Hitchcock, D.I., 160
Hittorf method for transference number measurements, 108–109
Höber, R., 165
Holzbauer, E.L.F., 243
Hückel, E., 128
Huggins, M.L., 47
Huxley, H.E., 229
Hydration of macromolecules, 14–15, 21
applied to spheres, 47
Hydrodynamics
and flow through capillary tube, 25, 27–31
and movement in concentric shells, 106
and Newton's law of flow, 24–25, 28, 29, 31, 119
application to spherical particles, 45
and non-Newtonian behavior, 25–30, 50–51
in obstructed systems, 52
Hydrogen electrodes
in cell without liquid junction, 151
in concentration cells
with liquid junction, 159
without transference, 156–158
in Daniell cell, 149
Hydrogen ions
binding by tobacco mosaic virus protein, 199, 210–212
movement of, and electron transport, 245
titration of, 139–140
transport with lactose, 181
Hydrolysis of ATP, 169, 173
catalyzed by dynein, 238
and dissociation of actomyosin, 228–229
thermodynamics of, 181–183
Hydrostatic pressure affecting equilibrium constant, 192–193

Ideal gas law, 92
Ideality of ions, departure from, 101–102
Interfaces, potentials at, 147–172
Intermediate filaments in cytoplasm, 236
Ion(s)
concentration-dependent interaction of, 101
conductances of, 104
diffusion rates of fast and slow ions, 80
exchange with absence of membrane, 161
hydration of, 110
ideality of, departure from, 101–102
mobility of, 108–111
transference number of, 104–105
measurement of, 108–109
unequal concentrations inside and outside of cells, 168–170
Ionic atmosphere, 94, 116
counter charges in, 118–119
and velocity of ions, 107
Ionic strength
in Debye-Hückel theory, 96
effect on mobility, 140

and polymerization of tobacco mosaic virus protein, 202, 207–208
Ionized electrolytes, osmotic pressure of, 5

Jaenicke, R., 14, 197
Johnson, K.A., 243

Karyokinesis, 237–238
Katchalsky, A., 82
Kauzmann, W.J., 49
Kirkwood, J.G., 139, 146
Kitzinger, C., 182
Klingenberg, M., 185
Klug, A., 197
Kraemer, E.O., 46
Krasny-Ergen, W., 51

Lactose transport, hydrogen ions with, 181
Lauffer, M.A., 14, 21, 28, 48–49, 53, 64, 125, 133, 195–198, 201, 204, 207, 208, 211–213, 220, 234–239
Le Chatelier principle, 192
Lee, B.K., 139
Levintow, L., 182
Lijklema, J., 99
Linderstrøm-Lang, K., 140
  dilatometer of, 194
Ling, G.N., 170
Liquid flow. *See* Hydrodynamics
Liquid junction potentials, 149, 150, 152–156, 171–172
  calculated values for, 162–163
  in concentration cells, 158–161
  and salt bridges, 126
  and transport in absence of membranes, 161
Loeb, J., 166
Longsworth, L.G., 67, 139

MacInnes, D.A., 92, 107
Macnab, R.M., 246
Macroions, diffusion rate of, 80–81
Macromolecules
  charged
    and elecroviscous effect, 51
    and equation for osmotic pressure, 16
    hydration of, 14
    and osmotic pressure in solutions, 9–13
  transport across membranes, 186–187
Magnesium, and hydrolysis of ATP, 182, 183
Markham, R., 198
McMichael, J.C., 213
Meister, A., 182
Melanophores, pigment migration in, 239
Membrane potentials, 163–166
  at interfaces, 147–172
  resting, 167–170
Membranes, transport across, 173–187. *See also* Transport across membranes
Meter-kilogram-second system, compared to c.g.s. system, 89—90, 91
Michaelis-Menten plots, and facilitated diffusion, 179
Microfilaments in cytoplasm, 234
Microtubules, 234
  in cell division, 237–238
  in melanophores, 239
  sliding of, and cilia motion, 238, 243–244
Miller, G.L., 49
Millikan, R.A., 38
Mita, D.G., 81–82
m.k.s. system compared to c.g.s. system, 89–90, 91
Molality
  and activity coefficient of electrolytes, 99, 101
  effective, 15
  and osmotic pressure, 2, 5
Molarity, and diffusion-limited reaction rate, 73–74
Molecular weight, and osmotic pressure, 16–18
Moore, W., 92
Morales, M.F., 182
Moseley, H.M., 82
Moyer, L.S., 116
Müller, H., 102
Muscle contraction, 224–234
  and conversion of chemical energy to mechanical work, 232–233
  interaction of myosin and actin in, 227–229
    regulation by troponin and tropomyosin, 233

mechanics of, 229–234
sliding filament-rotating cross-bridge model of, 229–230
Myofibrils, 224–226
Myosin filaments, 224–226
cross-bridges with actin, 229–230
interaction with actin, 227–229, 243–244
in cell division, 239
in cytoplasmic streaming, 240
equilibrium constant in, 227–228, 247
regulation by troponin and tropomyosin, 233
polarity of, 227

Nernst equation, 149
Neutrality, electrical
and activity coefficient of electrolytes, 103, 114
and diffusion from concentrated to dilute solution, 160
and diffusion of salts, 111
and electrochemical reactions in solutions, 108
and liquid junction potentials, 156
and streaming potential, 124
Newton, R., 15, 21
Newtonian liquids, 25, 30
Newton's law of flow, 24–25, 28, 29, 31, 45, 119
application to spherical particles, 45
Nexin, and cilia movement, 242
Nollet, Abbe, 1
Non-Newtonian systems, 25, 30, 50–51
Northrop-Anson cell, 63

Obstructed systems
diffusing particles in, 79
resistance in, 51–53, 79
Ohm's law, 81, 103, 123
Onsager equation, 108
Opposite charges
distribution in ionic atmosphere, 94, 118–119
electrostatic center of, 108
in Helmholtz double layer, 94, 116
movement in electrophoresis, 128
and streaming potential, 122
Osmosis
anomalous, 173–177
mechanisms in, 175–177
negative, 173
and pore types in membrane, 174–175
positive, 173
electroosmosis, 115, 116–121
Osmotic pressure, 1–21
activity coefficients in, 14, 19
definition of, 1
and diffusion, 2, 59
of charged macromolecules, 81
thermal, 82
and Donnan equilibrium, 9–13
enthalpic interaction in, 8, 15–16
entropy changes in, 6–8
and excluded volume, 5–9
and hydration of macromolecules, 14–15, 21
ions affecting, 99
molality affecting, 2, 5
molecular weight affecting, 16–18
in solutions of charged macromolecules, 9–13
thermodynamic relationships in, 2–5
total equation for, 15–18
Oswald viscometer, 30
Overbeek, J.Th.G., 99, 129, 130, 138, 139

Partial pressure of vapor, 99–101
Permeases, affecting transport, 184–186
Perrin, F.J., 39
Pfeffer, W.F.P., 1
pH
and ATP hydrolysis, 182
and electroviscous effect, 51
and hydrogen ion binding by tobacco mosaic virus protein, 200–201, 209–211
and polymerization of tobacco mosaic virus protein, 195, 197, 198, 201
and proton motive force, 244–245
Phagocytosis, 187
Phase boundary potentials, 165
Phenol, separating sodium chloride and potassium chloride, 165
Phillips, R.C., 182, 183
Phosphate bonds, high-energy, in ATP, 181
Pigment migration in killifish, 239
Pinocytosis, 187
Podolsky, R., 182

Poise, as unit of viscosity, 25
Poiseuille, J.-L.M., 25–26
Poiseuille's law, 28, 30, 125
Poisson's equation, 92–94, 120
Polins in membranes, function of, 185
Polymerization
of protein subunits, 194–195
with tobacco mosaic virus protein, 195–215
of structural elements in cytoplasm, 234
Pores in membranes
diffusion through, 62–64
formation and permeability of, 185
gated, 186
specificity of, 185
types in anomalous osmosis, 174–175
Porter, K.R., 239
Potassium
concentations inside and outside of cells, 168–169
sodium-potassium pump, 169–170, 186
transport by valinomycin, 184–185
Potassium chloride solutions as salt bridges, 162–163
Potential energy, in Boltzmann distribution law, 92
Potential gardient
affecting shear, 121
and diffusion
of charged macromolecules, 80–81
of salts, 111
and passive transport, 173
Potentials at interfaces, 147–172
concentration cells
with liquid junction, 158–161
without transference, 156–158
Donnan, 166–167, 168
electrode, 147–152
liquid junction, 149, 150, 152–156, 171–172
membrane, 163–166
resting, in biomembranes, 167–170
salt bridges, 161–163
Potentiometer in Daniell cell, 148
Pressure force, 37
determination of, 54
and flow through capillary tube, 28–29
and shear, 35–37
Price, W.C., 49
Probability integral, in solution of Fick's second law, 69
Proteins
electrophoretic mobility of, 138–140
polymerization of subunits, 194–195
with tobacco mosaic virus protein, 195–215
titration curve of, 140, 145–146
Proton motive force, in bacterial motility, 245–246
Pseudopodia formation in amoebae, 236–237

Radially symmetrical diffusion
cylindrical, 74–79, 87–88
spherical, 70–74, 86–87
Ramachandran, A.B.V., 211
Random orientation, friction coefficient for, 39, 54–56
for cylinders, 138
Randomly oriented rods
as obstructing particles, 53, 65
and solution of Fick's second law, 86
Rayleigh, Lord, 53
Resistance, frictional, 23–58
and diffusion of charged macromolecules, 81
and electroviscous effect, 51
and flow through capillary tube, 25, 27–31
and friction coefficient of ellipsoids of revolution, 38–41
in non-Newtonian systems, 50–51
in obstructed systems, 51–53, 79
and rotational friction, 31–35
and Stokes' law, 23, 35–38
and viscosity of liquids, 23, 24–27
Resting potentials in biomembranes, 167–170
Retardation effect, electrophoretic, 105–107, 129
Retardation force, and flow through capillary tube, 29
Revolution, ellipsoids of, friction coefficient of, 38–41
Reynolds number, 30
Richards, F.M., 139
Ringer's solution, ionic strength of, 118
Robinson, J.R., 50
Roepke, R.R., 45

Rotational friction, 31–35

Salt bridges, 161–163
Salting-out, and polymerization of tobacco mosaic virus protein, 202, 207
Salts
  diffusion of, 111–114
  dissolved in water, and volume of solution, 110
Sarcolemma, 224
Sarcomeres, 226
Sarcoplasmic reticulum, 233
Satir, P., 243
Scatchard, G., 11, 164
Schachman, H.K., 49
Schantz, E.J., 53, 64
Schlessinger, B.S., 140
Sedimentation, and solution viscosity, 48–50
Sedimentation potential, 115
Shalaby, R.A., 164, 198, 199, 201, 204, 207–209, 211, 212, 220
Shear
  and flow of liquids, 24, 25, 27, 29
  potential gradient affecting, 121
  and pressure force, 35–37
  and rotational friction, 35
  and viscosity of liquids, 25, 41
Shearing force, 37
  determination of, 54
Silver-silver chloride electrode, in cell without liquid junction, 151–152
Simha, R., 47, 50
Skou, J.C., 169
Sliding filaments
  in dynein-microtubule system, 238, 243–244
  in myosin-filamentous actin system, 229–230, 243–244
Sodium concentrations inside and outside of cells, 169–170
Sodium-potassium ATPase, 169, 186
Sollner, K., 174
Soret coefficient, 82
Soret effect, 82
Southern bean mosaic virus, sedimentation studies of, 49
Spheres
  charge per unit area of, 130–132, 143
  equal semi-axes in, 39
  suspension of, viscosity in, 41–47, 58
  symmetrically charged, electric field strength outside of, 97
Spherical charged particles, and Debye-Hückel theory, 94–99
Spherical diffusion, radially symmetrical, 70–74, 86–87
Stauffer, H., 209
Steinberg, I.Z., 233
Stevens, C.L., 14, 195, 196
Stokes equation for rotational friction coefficient, 34
Stokes law, 23, 35–38, 42, 73, 106, 109, 110–111, 128
  verification of, 37–38
Streaming, cytoplasmic, 239–241
Streaming potential, 115, 121–126
  and electroosmotic flow, 125
Stubbs, G.J., 211, 215
Surface adsorption theory, and unequal ionic concentrations inside and outside of cells, 170
Symtransport, 181

Tanford, C., 28, 139, 146
Taylor, E.W., 229
Temperature
  affecting viscosity, 26, 27
  in Boltzmann distribution law, 92
  and Debye-Hückel constants for aqueous solutions, 102
  and Debye-Hückel theory, 96
  and equilibrium constant, 193–194
  and hydrogen ion binding by tobacco mosaic virus protein, 200–201
  and polymerization of tobacco mosaic virus protein, 195, 208
  and thermal diffusion, 82
  and transport across membranes, 184
Thermal diffusion, 81–83
Thermodynamics
  of ATP hydrolysis, 181–183
  of cytoplasmic streaming, 240
  and Donnan potentials, 166
  in entropy-driven processes, 189–195
  first law of, 189–191
  of myosin interaction with actin, 227
  second law of, 191
Time of relaxation effect, 108, 129, 130

Tobacco mosaic virus
ion binding by, 164
polymerization of protein in, 195–215
bovine serum albumin affecting, 212
calcium binding in, 213–214
double discs in, 198, 201, 212
double spiral in, 198, 201, 212
electrical work in, 201–202, 205–207, 220
factors affecting, 198–199, 201
in flavum and E66 strains, 207
glycine affecting, 212
glycylglycine affecting, 212
helical state in, 212
hydrogen ion binding in, 199, 200–201, 210–212
ionic strength affecting, 202, 207–208
isodesmic mechanism in, 197–198, 209
light-scatter studies of, 202–203, 207
pH affecting, 195, 197, 198, 201, 209–211
in ribgrass virus strain, 208
salting-out factor in, 202, 207
sedimentation pattern in, 205
temperature affecting, 195, 208
two modes of, 201
water release in, 197, 216–219, 221
weight change in, 196–197
sedimentation studies of, 48, 49
viscosity of, 50
Torque, and rotational friction, 32–33
Transference number of ions, 104–105
measurement of, 108–109
Transport across membranes, 173–187
active, 173, 183–184
carrier-mediated, 181
permease function in, 185–186
secondary, 181
anomalous osmosis in, 173–177
antitransport, 181
carrier-mediated, 178–179
active, 181
passive, 181
diffusion and facilitated diffusion in, 177–181
group translocation in, 184
macromolecules and particles in, 186–187
passive, 173, 179
carrier-mediated, 181
specficity of channels in, 185
permeases in, 184–186
symtransport, 181
temperature affecting, 184
Traube, M., 1
Treffers, H.P., 45
Tropomyosin, 224, 226, 233
Troponin, 224, 226, 233
Tubulin, 234
Turbulence, and velocity of flow, 30

Valinomycin, potassium transport by, 184–185
Vand, V., 45–46, 48, 52
Van't Hoff, J.H., 82
Van't Hoff enthalpy, in entropy-driven processes, 221
Van't Hoff equation, 5
Vapor, partial pressure of, 99–101
Velocity
of cytoplasmic streaming, 240
in flow through capillary tube, 29
and force of resistance, 37–38
of ions in concentric shells, 106–107
in Newton's law of flow, 24
and rotational friction, 31–35
in suspension of spheres, 41–42
and turbulence, 30
and viscosity coefficient, 30
Vinograd, J., 15, 21
Virial coefficients, 5
Virial equations, 5
Viscometer, Oswald, 30
Viscosity
determinations of, 30
in diffusion, 48
and electroviscous effect, 51
and fluidity, 45–46, 58
intrinsic, 46–47
of ellipsoids of revolution, 47–48
of liquids, 23, 24–27
in obstructed systems, 52
in sedimentation, 48–50
structural, 50–51
of suspension of spheres, 41–47, 58
temperature affecting, 26, 27

Volume, exclusion from. *See* Excluded volume
v. Smoluchowski, M.V., 72, 126

Wang, J.H., 53
Water
  equivalent conductance of electrolytes in, 107
  glycerin interaction with, 111
  in hydration of ions, 110
  in hydration of macromolecules, 14–15, 21
  release in polymerization of tobacco mosaic virus protein, 197, 216–219, 221
  viscosity of, 31
Wiersema, P.H., 130

Yasui, M., 227

Zinc electrode in Daniell cell, 147–148